A Manufactured Plague?

To my parents

A Manufactured Plague?

The History of Foot and Mouth Disease in Britain

Abigail Woods

London • Sterling, VA

First published by Earthscan in the UK and USA in 2004

ISBN: 1-84407-081-6   paperback
       1-84407-080-8   hardback

Typesetting by JS Typesetting Ltd, Wellingborough, Northants
Printed and bound in the UK by Cromwell Press Ltd, Trowbridge
Cover design by Ruth Bateson

For a full list of publications please contact:

Earthscan
8–12 Camden High Street
Tel: +44 (0)20 7387 8558
Fax: +44 (0)20 7387 8998
Email: earthinfo@earthscan.co.uk
Web: **www.earthscan.co.uk**

22883 Quicksilver Drive, Sterling, VA 20166-2012, USA

Earthscan publishes in association with WWF-UK and the International Institute
for Environment and Development

A catalogue record for this book is available from the British Library

Library of Congress Cataloging-in-Publication Data

A manufactured plague : the history of foot-and-mouth disease in Britain / by Abigail
Woods.
       p. cm.
    Includes bibliographical references and index.
    ISBN 1-84407-080-8 (hardback : alk.)
    1. Foot-and-mouth disease–Great Britain–History. I. Title.

SF793.W66 2004
636.089'458–dc22

                                                                        2004010130

This book is printed on elemental chlorine free paper

Contents

List of plates

(see plate section)

Acknowledgements

This book represents the fruits of four years' work, undertaken towards the degrees of MSc and PhD at the Centre for the History of Science, Technology and Medicine (CHSTM), Manchester University. I owe an immense debt of gratitude to the many people who helped me to research and write it, and to keep body and soul together in the meantime. On the academic side, I must thank, above all, my supervisor, Professor John Pickstone, without whom I would never have started or finished this project. My colleagues at CHSTM have also provided precious help and encouragement. In addition, I am indebted to Pete Atkins, John Fisher, Anne Hardy, Peter Koolmees, Alison Kraft, David Smith and Keir Waddington, whose related research greatly assisted my work. Also to Brigitte Nerlich and her colleagues working on foot and mouth disease (FMD) at the Institute for the Study of Genetics and Biorisk in Society (IGBiS) at Nottingham University. Furthermore, I must thank those individuals and institutions that allowed me the opportunity to present my work at seminars and conferences, and members of the audience for their questions and comments.

Many people provided invaluable assistance as I followed the paper trail between different libraries and archives. I wish to express my gratitude to the staff of the Public Records Office (the National Archives), the British Library, the John Rylands University Library, Manchester, the Harold Cohen Library at Liverpool University and Cheshire Records Office, Chester. I located many important documents and photographs thanks to Jonathan Brown and his staff at the Rural History Centre, University of Reading and to Deborah Walker and Frances Houston of the Royal Veterinary College Library. Tom Roper of the Royal College of Veterinary Surgeons' library was helpful and encouraging throughout, and I am very grateful to Dr Donaldson of the Institute for Animal Health (IAH), Pirbright, for allowing me to access the institute's archives.

I would also like to thank several retired veterinary surgeons for their help and assistance. First, Hubert Skinner, whose cataloguing of the IAH archives facilitated my research, and who gave generously of his time to tell me of his life and work at the institute. Second, I must thank Mary

Brancker and Professor Alisdair Steele-Bodger for relating their first-hand experiences of veterinary practice during the mid 20th century. I am also indebted to Howard Rees, who told me of his years in the State Veterinary Service, and to Keith Meldrum and Chris Shermbrucker who pointed me in the direction of valuable documents. Finally, I wish to thank all those people who shared their experiences of the 1967 FMD epidemic at the recent Wellcome Witness Seminar.

The reappearance of FMD in 2001 led to a flood of media interest in my work. John Vidal of the *Guardian* kindly commissioned a piece on the history of the disease, while journalists at the *Sunday Times* and the *Independent* helped to disseminate my findings. I wish to thank Lawrence Woodward of the Elm Farm Research Centre and members of the National Foot and Mouth Group for their encouragement; also Ian Hunter of Littoral Arts for involving me in the 'culture' of FMD.

I owe a debt to many friends, who offered accommodation during my treks to the archives, plied me with alcohol and reminded me that there were things in life other than FMD. In particular, I must thank Ali, Anneli and Dave, Tina and Dan, Jo and James, Rob and Ronnie, Faye, Karen, Esther and Dustin. I could not have survived without the moral support of my Manchester friends, Adrian, Helen, Marion, Francis, Sarah and Simone, Carsten and Aya. Nick arrived on the scene in good time, and, together with my parents, provided never-ending support and encouragement.

I must thank Phillip Johnston, former headmaster of Queen Elizabeth's Grammar School, Blackburn, for his energetic and enthusiastic reading of my manuscript. Last, but not least, I wish to express my gratitude to the Wellcome Trust for their generous and ongoing financial support.

List of Acronyms and Abbreviations

AHDI	Animal Health Division I (MAFF)
ARC	Agricultural Research Council
BSE	bovine spongiform encephalopathy
BT	Board of Trade
BVA	British Veterinary Association
COBRA	Cabinet Office Briefing Room
COSALFA	Comisión Sudamericana para la Lucha Contra La Fiebre Aftosa
CVO	Chief Veterinary Officer
DCVO	deputy chief veterinary officer
DEFRA	Department of the Environment, Food and Rural Affairs
EUFMD	European Commission on Foot and Mouth Disease
EEC	European Economic Community
EU	European Union
EUFMD	European Commission for the Control of Foot and Mouth Disease
FAO	United Nations Food and Agriculture Organization
FMD	foot and mouth disease
FMDRC	Foot and Mouth Disease Research Committee
FMDRC CM	Foot and Mouth Disease Research Committee Meeting
FMDRC CP	Foot and Mouth Disease Research Committee Paper
FO	Foreign Office
IAH	Institute for Animal Health
IGBiS	Institute for the Study of Genetics and Biorisk in Society
GDP	gross domestic product
IRA	Irish Republican Army
MAF	Ministry of Agriculture and Fisheries
MAFF	Ministry of Agriculture, Fisheries and Food
MI5	Ministry of Intelligence
MOD	Ministry of Defence
MOH	medical officer of health
MP	member of parliament

MRC	Medical Research Council
NIMR	National Institute for Medical Research
NFU	National Farmers' Union
NVMA	National Veterinary Medical Association
OIE	Office Internationale des Epizooties
Panaftosa	Pan-American Foot and Mouth Disease Bureau
PAHO	Pan-American Health Organization
RAS	Royal Agricultural Society
RCVS	Royal College of Veterinary Surgeons
RVC	Royal Veterinary College
RVO	regional veterinary officer
TB	tuberculosis
UN	United Nations
VLA	Veterinary Laboratories Agency
WHO	World Health Organization
WTO	World Trade Organization

One feels guilty insulting the dead,
Walking on graves. But this pig
Did not seem able to accuse.

It was too dead

(Ted Hughes, *View of a Pig*, 1960)

Introduction

Just what is foot and mouth disease (FMD)? Scientists say that it is a viral disease of cloven-footed mammals, which causes painful blisters to appear in mouths and on feet and udders, and leads to lameness, drooling at the mouth, a loss of appetite and reduced milk production. They explain that, in the short term, many diseased animals will suffer. Some, especially the weak and the young, will die, and even animals recovering from the disease will show costly long-term reductions in growth rates and milk production. They also point out that FMD is one of the most contagious diseases known to man and can spread in almost every way imaginable.

Farmers view FMD from a very different perspective. They say that it is one of the diseases they fear the most. It stops them from moving livestock around and sending them to market, restricts their social lives and leaves financial hardship in its wake. An unfortunate minority can recall the day when, with a heavy heart, they rang the vet to report suspicious symptoms among their stock. They tell of their pain at seeing their animals slaughtered, a lifetime's work destroyed; of the silence the next morning, and knowing that life would never be quite the same again.

For the veterinary surgeon, FMD is not just any old disease, and infected animals are not patients like any other. Those who work for the Ministry of Agriculture tell of the enormous pressure to contain FMD as quickly as possible, and of the many difficulties involved in tracing its spread and stamping it out. They recall how, at times of crisis, they worked hour after hour, day after day, directing the destruction of animals that they had spent their careers trying to save from death and disease. And how, despite working to the best of their abilities, they frequently faced a critical public and a hostile media. Veterinary surgeons in practice speak of their feelings of helplessness as farming clients see their animals succumb to the disease, of the financial worries brought on by the loss of work, and of the anxiety that they might inadvertently transfer the virus between farms.

From the viewpoint of agricultural policy-makers, FMD presents a serious political problem. They speak of their extensive efforts to secure Britain's borders against the repeated invasions of this much-feared plague, and of the political rows that ensued both at home and abroad. They explain just how difficult it is to devise measures to satisfy everyone concerned, and complain about the many criticisms they have faced in the course of their work.

These are not the only people touched by FMD. We must not forget butchers, auctioneers, dealers, hauliers, importers and exporters of meat and livestock, participants in the tourist trade, members of rural communities and journalists. All have their own sad stories of FMD, stories that vary over space and time, and between regions, countries, continents and generations.

This history relates their experiences with FMD. And it explains why, over time, FMD came to mean what it did. For history was not inevitable. FMD did not have to become a feared disease. It did not necessarily have to be controlled, time after time for over a century, by isolating or slaughtering infected animals and restricting the domestic and international movements of meat and livestock. And scientific enquiries into the disease could have been pursued under very different circumstances, with very different results and policy implications. Although FMD is a 'natural' entity, which indisputably possesses certain biological characteristics, the reason why it is so feared, and the reason why its appearance causes such devastation, is because of the methods used to control it. Those methods are man-made. It is the thoughts and actions of humans that have 'manufactured' FMD into the particular political, economic, social and psychological problem that it is today. Had those thoughts and actions been different, then FMD might now pose a very different challenge.

This book sheds light upon the process by which FMD was manufactured. It reveals how, at different times and places, people have experienced and understood FMD, and why they responded to it in the ways that they did. It describes how policies were made, attacked and defended, and what concerns shaped scientific enquiries into the disease. Throughout, it shows that, had circumstances been different, and had the voices of some people carried more weight than others, FMD could have become a very different disease and had an impact far removed from that of today.

Some commentators will doubtless disagree with this tale. They will say that the dangers of FMD are self-evident and scientifically defined. They will claim that only the deluded or the ignorant ever disputed its status as one of the world's worst animal plagues, but that, thankfully, enlightened individuals managed to overcome this intransigence and educate the nation as to its true nature. They will also argue that the century-old methods of slaughtering infected animals, restricting livestock movements and limiting the import of goods from FMD-infected countries represented the only intelligent response to the disease. To back up their claims, they will point to the historic successes of this policy in keeping FMD out of Britain and in stamping out the outbreaks that did occur. They will also highlight the fact that all developed nations followed Britain's example in adopting it. *Ipso facto*, it had to be.

But history, when it is done properly, does not judge its subjects on the basis of present-day preconceptions; rather, it asks how and why, at different times and places, people thought and acted in the manner that they did. So let us suspend our modern conceptions and go back to the year 1839, when FMD first appeared in Britain. We will ask what FMD meant when understandings of disease, the state of agriculture, the nature of scientific enquiry and the role of government were all very different from today. Then we will move forward in time, assessing how and why reactions to FMD changed in relation to the political, economic, social, scientific and geographic context. And instead of concentrating on people whose opinions directly shaped the world as it is today, we will ask how everyone understood FMD and why. In this manner, we can restore to historical significance the voices of ordinary people, and those whose views deviated from the status quo.

This approach will reveal a history beset by controversy, in which different parties, at home and abroad, repeatedly clashed over the nature and control of FMD. It will demonstrate that currently incontrovertible facts were once nothing of the sort, and that 'enlightened' and 'ignorant' individuals alike acted, at least partly, on the basis of self-interest. Above all, it will show that while the UK government's long-term approach to FMD control was by no means irrational, it was not the only or even the most obvious method of tackling the disease, and that it prevailed not by any obvious superiority, but because its supporters wielded considerably more political and economic power than their critics.

Because so many different political, economic, cultural and social factors coloured ideas about FMD, and influenced reactions to the disease, the history of FMD in Britain is the history of Britain in microcosm. FMD helps to extend our knowledge of agriculture – for example, by revealing that 'modern' concerns about intensive farming and large-scale livestock movements are actually 150 years old. It provides insights into the historic development of the meat trade, international political and commercial relations, notions of national identity and fears of germ warfare. It illuminates the rise of the veterinary profession, changing understandings of disease and the development of 'modern' scientific ideas and practices. FMD has, at different times, shaped Britain's relationship with Ireland and Argentina. And through its history, we can learn how British veterinarians, doctors, cattle breeders, farmers, politicians, scientists, and meat and livestock traders lived, worked and related to each other.

What follows is a roughly chronological account, in which each chapter describes a particular theme, event or controversy, from 1839 to the present day. It is built upon an extensive range of previously unexamined material, much of it unpublished. Important sources include the National Archives at the Public Records Office, the archives of the Institute for Animal Health, Pirbright, records of national farming organizations, Hansard Parliamentary debates, Parliamentary papers, and national and local press reports. It is, above all, a history of politics, society and knowledge. It does not detail the scientific discoveries in the field of FMD research or the legislative responses to the disease; these histories have already been written.[1] And while more information upon the international FMD situation is desirable, it is, sadly, beyond the scope of this book.

As we all know, FMD returned with a vengeance to Britain in the spring of 2001. History was brought to life, and the ensuing epidemic was one of the most devastating on record. It is still too early to evaluate the long-term implications of that event, especially as many of the details will not come to light for another 30 years, when government files are declassified. But it is possible to assess the extent to which the events of 2001 were shaped by the past. This book will document the similarities and differences between that and previous epidemics and ask whether, in the light of history, the official response to FMD was justified.

For we are not, as the Anderson inquiry into the 2001 epidemic claimed, 'destined to repeat the mistakes of history'. Lessons can and should be learned. This account is written in the hope that readers will take heed of the past and use it to make sense of recent events and to plan their future response to FMD. The next time the disease appears – and there will inevitably be a next time – there will be no excuse for repeating the same century-old mistakes.

Chapter 1

Foot and Mouth Disease in 19th-century Britain: From Everyday Ailment to Animal Plague

FMD STRIKES

It was in August 1839, the second year of Queen Victoria's reign, that the owner of a large Islington dairy herd noticed that six of his cows were limping and had begun to drool. Perplexed by their illness, he did something that at the time was rather unusual – he summoned the vet. Mr Hill, veterinary surgeon, had learned his craft during a short course of study at the Royal Veterinary College (RVC) in London. Like the vast majority of his peers, he made a modest living attending to horses, and rarely visited sick cows, which were more usually treated by their owners or by lay healers known as cow leeches. However, in the fiercely competitive world of animal doctoring, he was grateful for any form of employment and hurried to attend the affected herd. After weighing up the evidence, he decided that the cows had probably eaten a poisonous plant while grazing at pasture.[1] He was soon proved wrong, however, as a flood of press reports revealed that the same disease had appeared in Smithfield market and was spreading rapidly through Norfolk, Essex and Scotland, causing:

> *a decline of appetite, a drooping of the ears, a drowsiness of the eyes, and a grating of the teeth. Then the tongue, nose and mouth become blistered, the hoof separates from the skin, a violent discharge from the nostrils takes place, a powerful influenza seizes the animal, and then comes a great falling off in the flesh.*[2]

None remembered ever having seen this disease before. They called it 'murrain' (a generic term meaning plague), 'epizootic aphtha' and 'the vesicular epizootic' (or 'tic' for short), an 'epizootic' being the animal equivalent of an epidemic. Its most popular name, however, was that by which it is still known today: 'foot and mouth disease' (FMD).

Reports of FMD quickly attracted the attention of the Royal Agricultural Society (RAS), a new organization for wealthy landowners interested in advancing agriculture by the application of science.[3] In the first systematic attempt to find out about the new disease, it sent circulars to members, asking them of their experiences. It also consulted Charles Sewell, RVC principal, about disease management. A summary of the 700 replies appeared in the society's journal. Most correspondents agreed that FMD was highly contagious, but noted that infected animals quickly recovered and lost only 5 per cent of their value. Dairy cattle, however, lost up to 30 per cent in value because 'udders subsequently became inflamed and tumefied; and abscesses were formed, terminating frequently in a total loss of milk'. According to Sewell, sick animals should be bled and purged. He described how to make up and administer medicines, how to treat mouth and foot lesions, and recommended that farmers isolate the diseased to prevent infection from spreading.[4] However, most ignored his advice – some because they believed FMD beneficial, in that animals seemed to undergo a growth spurt on recovery, others because they felt that isolation only delayed the inevitable spread of infection and that it was best to 'get it over with'. They fed soft food to affected animals and dosed them with traditional family medicines and drenches made with Epsom salts or bicarbonate of soda. Those that appeared unlikely to recover they dispatched to the butcher's, where most diseased mid 19th-century livestock ended their days.[5]

FMD continued to rage during 1841; but farmers and dealers took little notice and continued to move livestock around and take them to market. According to veterinary surgeon George Brown, infected cows and pigs were displayed so frequently at Smithfield market that floor sweepers filled baskets with their shed hooves. Towards the end of the year, disease spread diminished and symptoms became less severe; but FMD never disappeared completely, and during the years 1845, 1849–1852, 1861–1863 and 1865–1866, waves of disease washed across the nation.[6] Writing in 1848, livestock owner Hall Keary claimed that FMD was now 'so universal that, like the measles or whooping-cough in the human subject, all cattle are expected to have it at least once in their lives'. The 1853 edition of Clater's *Cattle Doctor* agreed, and noted that the disease had 'scarcely spar[ed] a single parish'.[7]

Because FMD was so prevalent, farmers began to view it as an unavoidable fact of life.[8] This attitude, when viewed from a present-day perspective, seems incredible. Why were they not concerned by the rampant spread of FMD and its damaging effects upon meat and milk production? And why did they not try to avoid it, or ask the state for assistance? One popular explanation holds that mid 19th-century participants in the meat and livestock trade were simply uninformed, and that once they became educated as to the true nature of FMD they willingly accepted proposals for its legislative control.[9] This argument has appeal as it suggests that we are far more intelligent and progressive than our forebears; but it is simply not true. Early Victorian responses to FMD were different from ours, but they were not irrational. Judged in their own context, they made very good sense.

Mid 19th-century farmers were probably correct to see FMD as a mild ailment. It is now known that the infection causes more serious symptoms in highly bred and highly productive animals than in the traditional types of stock that populated most of Britain's farms 150 years ago.[10] Also, the economics of agriculture were very different back then. Most farmers were far less concerned with profit margins and productivity than their later 20th- and 21st-century counterparts, and kept cattle not to make money, but to prop up the mixed farming system under which cows produced dung to fertilize arable crops and fed off home-produced roots and grain. Moreover, as a curable, transient ailment, FMD had little impact in an age when disease was so prevalent that most livestock existed in a suboptimal state of health.[11] As one commentator noted:

> *Health? What does that matter to an animal which has to live but a year or two? I grant you that it would die naturally before its time and that it is a race between the butcher's knife and the dart of death; but the knife is quicker on its pins.*[12]

Although 'germs' in the modern-day sense were scarcely known, mid 19th-century commentators had a sophisticated understanding of disease. They blamed most illnesses upon 'morbid poisons' that generated spontaneously and spread by contagion. They believed that the symptoms of disease and its capacity to spread were influenced by the state of the atmosphere, the presence of dirt and damp, and the patient's hereditary constitution and nutritional or emotional state.[13] Veterinarians and farmers understood FMD within this model. In as much as it spread rapidly within and between herds, it possessed a highly contagious nature; but when it appeared in herds and flocks that had not recently mixed with other animals, its generation was 'spontaneous'. They felt that while quarantine-type measures could limit the spread of contagious disease, it was more difficult to anticipate and prevent spontaneous disease generation; and in the belief that FMD could not be avoided, they grew to accept its appearance.[14]

One question commonly asked was where FMD had come from. Later 19th-century commentators blamed the importation of diseased livestock from the Continent, an explanation that is generally accepted today.[15] However, at the time, it was not so credible because livestock imports were officially banned until the mid 1840s, when the repeal of the Corn Laws opened British ports to foreign agricultural produce and ushered in an era of free trade. To most contemporary observers, the appearance and spread of FMD was the direct result of changes to animal husbandry and the livestock trade. Mid 19th-century Britain was in the throes of the Industrial Revolution, and as rural inhabitants flooded into the towns in search of work, so the urban demand for meat rose. A network of cattle dealers and traders sprang up to supply this new market. Thanks to railway and steamship development and the opening of the ports, they were able to move large numbers of livestock quicker and over far greater distances than ever before. Animals in transit were packed together closely in poorly ventilated trucks or holds and deprived of food and water. These were the very conditions believed responsible for the formation and spread of infection, so few were surprised to discover that animals commonly arrived at their destinations suffering from FMD.

For their milk, city-dwellers turned to urban dairies. The owners of these dirty and poorly ventilated establishments bought in newly calved cattle from the country and fed them on brewer's grains (an 'unnatural food') until their milk dried up and they became butchers' meat. FMD often put in an appearance at the dairies, which were well-known 'hotbeds of disease'. Meanwhile, agricultural practices were changing as numerous farmers stopped feeding their animals grass or hay during the winter, instead using artificial feeds made of linseed, rapeseed or cotton seed. Critics believed that this type of fodder increased the body's susceptibility to diseases in general, and FMD in particular.[16]

As for controlling FMD, few mid 19th-century British farmers would have expected or welcomed state intervention. While, today, central government governs many different aspects of animal and human health, during the 18th and early 19th centuries it was mostly concerned with tax collection and national defence, and left other initiatives to private individuals, charities or local authorities. Only during major epidemics did it intervene, and then only to enforce quarantine, a measure first applied during the 16th-century plague epidemics; ships arriving from plague- or yellow fever-infected ports were also routinely isolated. Quarantine was extremely unpopular because it halted trade and prevented people from fleeing disease.[17] The Privy Council meted out similar treatment to animals suffering from rinderpest or cattle plague, an extremely fatal and highly contagious ailment that appeared in Britain in 1714–1715, 1745–1758, 1769, 1774 and 1781. It banned livestock movements, required farmers to isolate sick cattle and offered compensation if they slaughtered diseased stock. Livestock owners resisted and evaded these measures, and local justices of the peace who sympathized with their plight rarely attempted to enforce the law. Nevertheless, cattle plague eventually disappeared and did not return for almost a century.[18]

Central government's interest in public health grew during the mid 19th-century as industrialization gave rise to the tremendous social and environmental problems so vividly depicted by the novelists Charles Dickens and Elizabeth Gaskell, and the social commentator Frederick Engels in his 1844 book *The Condition of the Working-class in England*. The population of England and Wales grew from 9 to 18 million between 1801 and 1850, and as towns grew ever dirtier and more crowded, paupers died in their thousands from disease and deprivation. Parliament responded to the growing crisis by passing new public health legislation; but in accordance with the political ideology of the day – which emphasized state nonintervention (known as *laissez-faire*) and self-help – most new laws were permissive and merely enabled the local authorities to introduce appropriate measures if they so wished.

At about the same time, resistance to quarantine grew, partly because it curtailed individual liberty (an attribute that many saw as a natural right of all British subjects), and partly because it was seen to fail during the 1832 and 1848 British cholera epidemics. The cholera experience contributed to a growing feeling among doctors that the spontaneous generation of disease was more important than they had once thought. Consequently, quarantine – a measure that could only help to control contagious diseases – ceased to make sense. Another reason why its popularity nose-dived was that the quarantining of ships contravened free trade, a policy that grew to virtually constitutional status on account of the tremendous growth in the British economy that followed its mid 19th-century institution.[19] At times like these, few British farmers would have wanted or dared to suggest that the government control FMD in the same manner as plague, cholera or cattle plague.

However, minor legislative FMD controls crept in by default in 1848 as a result of government efforts to control sheep-pox, a contagious and highly fatal disease that had been imported with Spanish sheep the year before. Sheep-pox was believed to be a fitting subject for legislative control because of the precedent set by its human equivalent, small-pox. Using an eight-year-old boy as an experimental subject, Edward Jenner had demonstrated in 1796 that inoculation with the cow-pox virus protected humans from small-pox infection. The practice soon took off. Acts passed in 1840–1841 made free vaccination universally available, and in 1853 it became compulsory for all infants.[20] The success of this method – which was not without its critics – meant that when the previously unknown sheep-pox appeared in 1847, inoculation was one of the first control methods investigated by veterinary surgeon J B Simonds, holder of a new RAS-sponsored chair in cattle pathology at the RVC. Simonds believed that sheep-pox was, like its human equivalent, incapable of spontaneous generation and, hence, controllable by quarantine. He went on to advise the government on the passage of two Acts to limit the importation and movement of infected sheep.[21] In subsequent years the wording of this legislation gave rise to some confusion: did it apply only to sheep-pox, or to other contagious animal diseases such as FMD and bovine pleuro-pneumonia (another new disease which first appeared in 1842)? In practice, one Act, which prevented the movement and sale of infected animals within Britain, was applied only to sheep-pox. The other Act, which provided for the inspection of livestock imports, was applied to a range of diseases so that when customs inspectors discovered a single imported animal suffering from FMD they would order the immediate slaughter of the entire cargo, much to the chagrin of livestock importers.[22]

JOHN GAMGEE AND THE DISEASED MEAT PROBLEM

It was not until the 1860s, almost 30 years after FMD first appeared in Britain, that a handful of individuals began to argue against the popular belief that it was a mild and uncontrollable disease. Chief among them was John Gamgee (1830–1894), a charismatic, ambitious and outspoken veterinary surgeon, who declared FMD an extremely costly and serious problem that was in need of a legislative solution. The son of a prosperous Florentine horse doctor and brother of two well-known doctors, Arthur and Joseph, John Gamgee belonged to a higher social class than most veterin-arians. He had also received a better education, having toured veterinary schools on the Continent, where training was more scientific and veterinary surgeons were held in higher regard than in Britain. In 1856 he took up a post at the Edinburgh veterinary school, but quickly fell out with his employer, the aging Professor Dick, and left to establish a rival 'New Edinburgh Veterinary College', which boasted a Continental-style veterinary education.[23]

During the late 1850s and early 1860s, Gamgee and his brother Joseph waged a private war against the sale of diseased meat. At that time, there was rising concern over the deteriorating quality of food, especially the food of the urban working classes, which was commonly adulterated by greedy shopkeepers. Like most other trades, the buying, selling and handling of food was unregulated; but with the development of new chemical analytical techniques, it became possible to detect adulteration, and the published results of several investigations resulted in widespread demands for official action.[24] One of the most influential inquiries was carried out by the Analytical Sanitary Commission, set up in 1850 by Thomas Wakley, editor of the medical journal the *Lancet*. For several years he and his medical colleagues collected samples of food and subjected them to microscopic and chemical analysis. The *Lancet* published the names and addresses of offending retailers, among them shopkeepers who had sold milk contaminated with sheep's brains or coffee adulterated with earth and chicory. It also pressed for legislative action on the grounds that adulteration amounted to fraud and threatened the public's health. The government's 1855–1856 Select Committee on Food Adulteration investigated the nature and scope of the problem, and new legislation soon followed in 1860. This made it illegal to knowingly sell adulterated food, and permitted local authorities to appoint public analysts to check samples. Few did so; but the Act, nevertheless, helped to reduce the scale of the problem.[25]

The 1860 Act did not directly address the sale of diseased meat, despite witnesses' complaints about the trade in:

> *slinked beef. . .a class of meat obtained from cows which are diseased and unfit for human food. The animals are subject to various diseases. Some are called ticked* [they have the epizootic FMD]; *some have the milk fever; some have worm i'th tail; some are gaped; others are broken-up old cows.*[26]

Some controls already existed under the 1855 Nuisances Removal Act, which permitted local authority inspectors to enter premises, inspect meat and seize any that proved unfit for human consumption.[27] Unfortunately, such measures were poorly enforced. Most inspectors were butchers who refused to report their erring colleagues, and farmers openly admitted that they could not make a living without selling diseased and even dead animals for human consumption.[28] In 1857 John Simon, president of the Medical Department of the Privy Council, commissioned Dr E Headlam Greenhow, lecturer on public health at St Thomas's Hospital in London, to report upon livestock disease, the sale of diseased meat and the effects of its consumption upon human health.[29] At about the same time, the Gamgee brothers began to visit Edinburgh dairies, markets and slaughterhouses. In a stream of letters to prominent politicians and to the national, local and medical press, they described the horrors of the diseased meat trade and issued demands for its legislative reform. John Gamgee also presented Simon with a voluminous report on the matter, in which he argued that a third of all meat sold was damaging to the public's health.[30]

Many livestock owners, dealers and butchers were downright hostile to Gamgee's claims, and regarded him as an irritating nuisance who had no right to interfere with their (largely legitimate) business practices. Public health doctors were more enthusiastic, and several voiced support for his campaign, though not the influential John Simon, who argued that as human health had not markedly deteriorated of late, diseased meat consumption could neither be as prevalent nor as dangerous as Gamgee had alleged.[31] Some historians regard Gamgee as a hero, an enlightened and far-sighted man who sadly failed to stir an ignorant and intransigent establishment.[32] But although, in later years, many of Gamgee's recommendations were implemented,[33] there were many good reasons why his campaign initially failed. The sale of diseased meat was only one of a number of pressing public health issues, and in contrast to food adulteration, there were no accepted methods of proving that meat was unfit for human consumption. The *laissez-faire* state frowned upon unnecessary trade regulation; and, at a time when veterinary surgeons had little status in the eyes of the public, Gamgee's opinions carried little weight within the political sphere.

Unperturbed, Gamgee tried to achieve his goal by different means. Instead of demanding direct restrictions upon the sale of diseased meat, he tried to stop livestock from becoming sick in the first place by campaigning for new measures to prevent contagious animal disease spread. Using statistics provided by bankrupt livestock insurance companies and Edinburgh dairymen, he calculated that livestock disease cost the nation UK£8 million annually, and deduced that because FMD caused a long-term reduction in meat and milk production, its associated losses outweighed those caused by any other disease. He argued that FMD, along with several other contagious animal diseases such as pleuro-pneumonia, arose not by spontaneous generation but by importation, and spread by contagion alone. They could easily be prevented if, under veterinary supervision, affected animals were slaughtered and livestock imports regulated.[34]

Gamgee tried to persuade his fellow veterinary surgeons to back these proposals. He told them to go out and educate ignorant farmers and the state about disease control, and promised that professional benefits would follow, as on the Continent where state-employed veterinarians had risen in terms of income and social status.[35] Unfortunately, he met with little success. Most veterinarians disagreed with his understanding of disease, and because their training extended only to horse doctoring, they knew far less about cows than most farmers. Also, Gamgee was not popular in the profession.[36] He was bombastic, boastful, and clashed repeatedly with Britain's leading veterinary expert on livestock disease, J B Simonds, whom he regarded as an ignoramus. On one occasion he went so far as to dig up some dead sheep to prove that Simonds's disease diagnosis had been incorrect; but he failed to dislodge or discredit his opponent, an austere man who refused to rise to the bait.[37]

Gamgee was rather more successful in drumming up the support of leading agriculturalists. Several were attracted by his claim that they would benefit financially from animal disease control. They also concurred with his portrayal of FMD as a severe disease requiring legislative control. Compared to the common stock, their well-maintained herds of valuable, highly bred pedigree cattle suffered severely from FMD, and the occasional fatalities and delays in breeding that it caused proved extremely costly. Also, because their standards of husbandry were high, they sensed aspects of the disease which ordinary farmers did not, notably its long-term effects upon meat and milk production. In 1863, RAS member Edward Holland MP introduced a private members' bill based upon Gamgee's proposals for livestock disease control, but withdrew it on learning that the government was planning its own legislation.[38] The following year, Parliament sat to consider a new Cattle Diseases Prevention Bill. This proposed measures for the control of FMD, pleuro-pneumonia, glanders (a rare but fatal horse disease that could spread to humans) and cattle plague (which raged on the Continent but was not present in Britain), to be executed by veterinary inspectors appointed by local authorities.[39]

Unfortunately for Gamgee, many participants in the meat and live-stock trade vehemently opposed the government's bill and fought long and hard to defeat it (see Plate 2). Its proposed restrictions upon the movement of diseased stock threatened their businesses, and they were outraged at the inclusion of FMD, which they thought a harmless and unpreventable illness. For veterinary advice they turned to Simonds, who made no secret of his opposition to Gamgee's proposals. They held public meetings condemning the bill, advertised its faults in the farming press and sent skilled witnesses to give evidence before a government select committee.[40] Feelings ran high, and at one committee session 'the Professor had rather a busy time of it whenever the room was cleared, as the opposition mustered strong and (two of them at first not in the most civil way) made a succession of little runs at him'.[41] Long before its hearings were complete, the committee decided to strike FMD from the bill. Shortly afterwards, John Hall Maxwell, secretary of the influential Highland Agricultural

Society, rounded up the Edinburgh dairy men whom Gamgee had named as sources for his statistics and persuaded them to sign a disclaimer stating that they had supplied no such information. This spelt the end for Gamgee, as the *Illustrated London News* reported:

> As the inquiry proceeds, it seems to be more and more evident that the very persons whose interests it is the avowed object of the legislature to protect are really those most opposed to the bill. They want to be left to conduct their own business, at their own risk, without any such official dry-nursing, and they regard the great army of locusts which would overrun the land in the shape of inspectors as an utter nuisance and delusion. They contend that the whole question has been blown into the most undue dimensions, and that it has collapsed into nothing now that evidence has been heard in reply... The Evil, as far as it exists, seems to them to be self-preventative. Men dare not, for fear of the future consequences to their business, bring beasts and sheep to market when their diseased condition is patent. . .they consider that an army of inspectors, such as Professor Gamgee might propose to educate, would be physically powerless at Falkirk and Ballinasloe [fairs], and practically powerless for anything but annoyance in a smaller sphere of action.[42]

On the advice of the select committee, the government withdrew its bill. Celebrating opponents claimed that the whole affair had 'taught the good old English lesson that individual effort was better than government assistance'.[43] But as we shall see, the meat and livestock trade did not long escape the threat of state interference.

THE CATTLE PLAGUE EPIDEMIC, 1865–1867

Britain in the year 1865 suffered from two devastating disease epidemics. The first was cholera, which appeared in September and killed 15,000 people before disappearing a year later. The second was rinderpest or cattle plague, a highly fatal and contagious disease that remained in Britain for over two years and caused the deaths of at least 420,000 cows, or 7 per cent of the national herd. A century had passed since Britain last experienced the cattle plague, and the quarantine-based measures that had once controlled it were long forgotten. As British farmers, veterinary surgeons and officials wrestled anew with the disease, important differences of opinion emerged. Gamgee and Simonds both knew that it had raged for some years on the Continent and deduced that it must have been carried into Britain by infected livestock imports. They also knew that the Continental search for a scientific cure or preventative had failed, and that the only practicable method of controlling disease spread was to restrict livestock movements and slaughter cattle exposed to infection. However, almost everyone else believed that both cattle plague and cholera had generated spontaneously as a result of atmospheric and environmental influences, and thought Gamgee's and Simonds's proposals objectionable and unworkable.

At first, the government tried to control cattle plague using the 1848 Sheep-pox Act. It asked local authorities to appoint veterinary inspectors who should monitor cattle markets for the presence of disease, restrict the movements of infected animals and slaughter them where necessary. They also gathered statistics of disease incidence and forwarded them to a newly established, temporary State Veterinary Department, headed by Simonds, with George Brown, formerly professor of veterinary science at the Royal Agricultural College, as deputy (see Plates 3 and 4). These measures failed to check the spread of disease, and as cattle died in their thousands, panic spread across the nation. A Royal Commission reporting in 1866 echoed Simonds's and Gamgee's recommendations, and ordered medical scientists to pursue scientific investigations. When they failed to find a cure or vaccine, the government stepped in. Reluctantly, it passed new legislation that led to the slaughter of all infected animals and their contacts. Railway transit was stopped, fairs and markets closed, and all livestock imports were slaughtered on disembarkation. A marked drop in disease incidence followed, although it was some months before the plague disappeared from Britain.

The cattle plague epidemic was a tremendously significant event that brought about major shifts in the understanding of livestock diseases, and led to the introduction of administrative machinery and disease control methods that still exist today. It came as an enormous shock to participants in the meat and livestock trade, who had hitherto viewed animal disease as a manageable occupational hazard, and its eventual elimination forced them to accept Gamgee's and Simonds's claims that it had arisen by importation, spread by contagion, and should be subject to legislative control.[44] Looking back, there was another reason for contemplation. It seemed that the new control measures had not only worked against cattle plague; they had also reduced the spread of FMD and pleuro-pneumonia. This finding suggested that, in contrast to previous opinion, the latter two diseases also spread mostly by contagion, arose rarely if at all by spontaneous generation, and could be controlled by legislative means.[45] However, the cattle plague experience did little to shake the popular belief that FMD was a mild, transient infection. Consequently, although the government decided to include FMD (and pleuro-pneumonia) in its 1869 Contagious Diseases of Animals Bill, it laid down far more lenient controls for FMD than for cattle plague: owners were not required to notify the authorities of the appearance of FMD; inspectors had no right of entry onto private property; and infected animals were not slaughtered but isolated, and banned from markets and fairs.[46] Parliament accepted these proposals and the bill became law. As a result, the 'temporary' State Veterinary Department gained new areas of responsibility and became a permanent body. It still exists today.

Unfortunately, the new Act had virtually no impact upon the extension of a new FMD epidemic, which emerged during the autumn and winter of 1869. Legislators then faced a difficult decision. Should they allow FMD control to pass back into the hands of the farmer? Or should they make a further attempt to contain the disease, this time using stricter, more wide-ranging regulations?[47] None could agree. For the next 15 years, the issue was debated repeatedly in Parliament, at veterinary, farming and trade association meetings, and by three government select committees of inquiry, but to no avail. The cause of FMD, the symptoms it produced, and the means by which it spread remained moot points; consequently, there could be no consensus upon the type and stringency of measures needed to control it. Divisions of opinion hadn't mattered before, when it was up to the farmer to decide what, if anything, to do about the disease; but now the stakes were raised as legislation threatened to impact upon the personal and professional interests of veterinarians, politicians, farmers and participants in the meat and livestock trade. Factions formed between and within these groups as individuals fought for a personally advantageous solution to FMD that seemed sensible in the light of their private disease experiences. Consequently, FMD control became a highly political and emotive issue.

Prominent in these debates were Simonds – now the acknowledged leader of the British veterinary profession – and Gamgee, who became an increasingly marginal figure as time passed. Their disagreements continued well into the 1870s, and their contradictory views of FMD shaped public opinion. Gamgee continued to tell veterinary meetings, farming clubs and government select committees that FMD was an extremely serious disease that entered the nation by importation and would die a natural death if the livestock import trade was halted for several months. Ever inventive, he tried to discover a technique for preserving dead meat in the hope that dead meat imports could replace the disease-ridden live animal trade. His views were taken up by a group of aristocratic agriculturalists, pedigree cattle breeders and Tory MPs, who were keen to secure a prosperous future for rural Britain and to guard the health of their own animals, which, as mentioned earlier, suffered severe FMD symptoms. But although several of his proposals eventually gained legislative expression, Gamgee received little credit and died a frustrated man in 1894.

Simonds, who remained head of the veterinary department until his 1872 appointment as RVC principal, and Brown, who succeeded him, maintained that FMD was not a dangerous disease and that wide-ranging control measures were inappropriate. While acknowledging that FMD was occasionally imported from abroad, they blamed the domestic livestock trade for most cases of disease spread, and recommended that instead of halting imports, the government should employ veterinary inspectors to detect and isolate infected British livestock. Naturally, such proposals were popular among struggling veterinary surgeons. They were also adopted by Liberal and urban members of parliament (MPs), who opposed import controls partly because they violated free trade, and partly because they would restrict the meat supply, thereby reversing the recent trend towards cheaper meat and placing it out of reach of the working classes. Like most members of the Victorian upper classes, these politicians were haunted by the spectre of working-class revolt, and they feared that the poor – who wanted to eat more meat – would grew restive if its price increased. Moreover, according to the new science of nutrition, meat consumption was essential to muscle growth and energy levels. If workers failed to eat enough, their productivity levels – and capitalists' profits – would flounder. Import restriction could not, therefore, be permitted. Tory agriculturalists countered such objections by claiming that dead meat imports could easily replace the live trade, and that freedom from FMD would increase the productivity of British livestock and give farmers the confidence they needed to breed more animals. Their Liberal opponents were not convinced by this interpretation of the laws of supply and demand, and accused them of seeking higher prices for their own livestock at the expense of the urban poor.

Participants in the meat and livestock trade were also divided over what to do about FMD, and for good reason. Some livestock owners were particularly prone to the effects of FMD, either because they owned dairy and breeding stock – which suffered especially serious symptoms – or because their businesses relied upon the purchase of new animals that frequently introduced infection into their flocks and herds. They were keener to get rid of FMD than dealers – who could quickly offload infected animals – and farmers who bred their own replacements, or engaged in the rearing and fattening of livestock. However, most parties were prepared to support legislative measures as long as they posed no threat to their businesses. Those involved in the domestic livestock trade naturally expressed a preference for import controls, while participants in the overseas and import trade argued the case for domestic trade restrictions.[48]

FMD BECOMES A PLAGUE

Surprisingly, out of this storm of controversy, political, professional and popular opinion slowly began to coalesce around a new view of FMD. By the mid 1880s – almost half a century after FMD first appeared in Britain – most commentators no longer saw it as a minor unavoidable ailment, but as a foreign invading animal plague that could and should be controlled by far-reaching and extremely stringent legislative measures. At the same time, the framework of the 20th-century FMD control policy came into being. But why did these changes occur? How did the vocal opponents of the 1864 bill transmute into fervent supporters of government intervention, and how did the once disregarded FMD become a feared foreign plague?

One popular explanation holds that veterinarians, politicians and farmers belatedly realized that FMD was, in fact, an extremely dangerous and damaging disease that had to be controlled by means of import and domestic trade restrictions. But this is simply not true. As we have seen, the clinical manifestations of FMD were extremely variable, it impacted more upon some businesses than others, and proposals for its control had different implications for different sections of the meat and livestock trade. Consequently, there existed many rival opinions about the disease, none of which stood out as any more moral or rational than the rest. So it is simplistic to assume that the present-day image of FMD as a terrible animal plague emerged because it was obviously correct. In fact, a whole range of social, economic, scientific, political and legislative matters helped to transform FMD from inconsequential ailment to animal plague.

In mid to late Victorian Britain, upper-class aristocrats and landowners wielded a substantial degree of political and economic power, and their views were accorded far more weight than the opinions of those who worked for a living. Social hierarchies also existed within farming and the meat and livestock trade: pedigree cattle-breeding was a highly prestigious activity; mixed or 'high' farming carried certain kudos; but dealers, middlemen and jobbers garnered little respect. Faith in British superiority was such that the opinions of Irish graziers and foreign importers carried little weight, while veterinary surgeons, whose professional aspirations had yet to become reality, had only limited authority. Responsibility for shaping legislation lay with members of parliament and the civil service, whereas those who stood outside the corridors of power could only hope to influence policy if selected to give evidence before a departmental commit-tee of inquiry, or by amassing deputations and calling on their MPs.[49] Within this setting, upper-class pedigree breeders and Tory MPs were by far the most influential livestock owners, and for reasons already discussed, they thought FMD a foreign and extremely serious disease. But while important, their political influence cannot solely account for why FMD became a plague; after all, their 1864 attempt to force through Gamgee's unpopular legislation had actually failed.

The changing fate of British agriculture also influenced the status of FMD. During the 1860s, farming had prospered; but in late 1870s and early 1880s, it fell into depression as the building of railroads across the American prairies led to a rise in cheap grain imports, while inclement weather reduced domestic production. For many mixed farmers, arable profits no longer cloaked the financial status of livestock production, and so the losses inflicted by FMD upon meat and milk production became more apparent and economically significant. Some activists even contended that the effects of FMD had made farmers reluctant to invest in cattle instead of corn, and argued that it was the government's job to protect them from this terrible disease.[50] However, for many farmers, legislative restrictions upon the movement and marketing of animals would have compounded their financial difficulties; therefore, declining agricultural fortunes cannot in itself explain the transformation of FMD.

Debates over FMD coincided with news of Pasteur's and Koch's formulation of the germ theory of disease. So did the realization that infectious diseases were caused by microbes change attitudes towards FMD? The evidence would suggest not. Contrary to popular assumptions, the germ theory and new bacteriological methods of disease investigation did not cause an overnight transformation in the understanding of disease. Instead, they were adopted piecemeal over a 20-year period, in different ways by different groups, with debates upon their validity peaking around 1875. As a result of their cattle plague experience, most veterinarians readily accepted that contagious germs were responsible for disease. However, they were reluctant to rule out spontaneous generation completely because it provided the most plausible explanation for the first-ever case of FMD. Also, they had little time for scientific theorizing or for new laboratory-based methods of diagnosing and investigating disease. Simonds and Brown viewed knowledge of disease pathology and the nature of the disease entity irrelevant to the question of FMD control. They had long recognized it as a contagious problem, and cared little whether it was caused by a living microbe or by a chemical poison, fungi or ferment, as was previously believed.[51] In comparison, medical scientists were more interested in bacteriological research; but there were few jobs and little state funding available for investigations into animal diseases. During the later 1870s, the RAS asked Dr John Burdon-Sanderson, a medical scientist who had carried out experiments for the 1866 Cattle Plague Commission, to research into FMD; but due to cost restrictions and poor facilities, his investigations failed to bear fruit.[52]

It seems that the factor most responsible for transforming FMD from inconsequential ailment to foreign animal plague was the legislation used to control it. We tend to assume that the measures used to control disease are based upon an understanding of its clinical symptoms and means of spread – that knowledge comes first and action follows. But for FMD, the reverse was true. As we shall see, the very formulation and operation of legislative FMD controls during the years 1869–1884 had an important and unanticipated influence upon experiences and understandings of the disease. It became increasingly feared and eventually was seen as one of the world's worst animal plagues, a highly contagious and economically devastating ailment akin to the cattle plague. It was in this sense that FMD was 'manufactured'. FMD-as-plague wasn't always 'out there' in nature, awaiting discovery by enlightened individuals. It was a new creation, a by-product of the processes involved in its control. And as this new vision of FMD grew in strength, its social origins were gradually obscured, until it came to be viewed as an incontrovertible fact of nature.

When the 1869 Act failed, the government tried to improve its FMD controls by introducing additional measures under order. Local authority veterinary inspectors were permitted to enter private premises and, on diagnosing disease, to declare an 'infected place' of 1-mile radius within which livestock movements were prohibited for several weeks. Around half of the local authorities made use of their powers; but complaints abounded that the new controls caused unnecessary trade interference, and, in 1873, on the advice of a select committee, the Liberal government abandoned them.[53] Several months and a general election later, they were reinstated by the new Tory government, which proved more sympathetic to the demands of leading agriculturalists. Again, they failed to check FMD spread, and in 1877, the cattle plague briefly returned to Britain.[54] The government responded by appointing another two select committees to reconsider contagious animal disease control, and based its 1878 Contagious Diseases of Animals Bill upon their recommendations. Parliament readily agreed upon the bill's proposals for more stringent controls against the domestic spread of FMD. However, its proposal to ban livestock imports from infected nations proved highly contentions, and after several long and extremely tortuous debates on the matter, the government decided to admit foreign animals, subject to their immediate slaughter on arrival.[55] Yet again, the new measures failed to have the desired effect, and in 1884, the Liberal government reluctantly agreed to ban all livestock imports from nations infected with FMD, pleuro-pneumonia, sheep-pox or cattle plague. Few countries were free of all these ailments, and so the livestock import trade virtually ceased. Shortly afterwards, endemic FMD disappeared from Britain.[56]

These recurrent debates and policy changes made FMD control an extremely prominent and pressing issue, and enhanced its profile considerably. Participants in the meat and livestock trade naturally wanted to know whether proposed regulations were likely to succeed, how they would impact upon their daily lives, and whether their benefits would outweigh the costs incurred. Consequently, they stopped ignoring FMD and began to scrutinize its clinical symptoms, economic effects and epidemiological behaviour. They took note of, and awarded new significance to, everyday events such as the importation of infected stock from Ireland and abroad, and the movement and sale of diseased animals within Britain. They also began to notice that in addition to its acute symptoms, FMD caused chronic reductions in meat and milk production. Reflections such as these prompted many livestock owners to abandon the belief that FMD was a mild, incidental illness and to support its legislative control.

In the meantime, new government legislation made it possible to estimate more accurately the costs of infection. Between 1870 and 1872, all livestock owners had to inform the local authorities when FMD appeared among their stock. Statistics were collated by the veterinary department, and revealed that in the three months to March 1870, there were outbreaks upon a staggering 16,140 premises.[57] Farming and veterinary representatives took these figures and multiplied them by the loss in value suffered by FMD-infected cows, a much-disputed sum that varied between UK£1 and UK£8 a head for dairy cows alone. The result was a figure that approximated to the national costs of FMD infection. Some translated this sum into volumes of meat and milk lost, bringing the effects of FMD home to urban consumers, as well as to agricultural producers.[58] Their calculations helped to raise awareness of the impact of FMD and drummed up support for more effective controls.

Before the cattle plague, the unexpected appearance of FMD would often have been explained by reference to spontaneous generation. Now, however, its arrival among animals that had had no known contact with the diseased was blamed upon the indirect transfer of germs upon clothes, boots or vehicles. New legislative initiatives made these indirect routes of disease transmission more visible: in restricting the movements of infected stock they reduced the opportunities for direct disease spread. In 1881, Brown traced the resurgence of FMD to infected French cattle, which had been slaughtered on arrival in Britain and propagated the disease indirectly.[59] This event revealed the danger posed even by dead foreign animals, and helped to strengthen the case for complete import prohibition.

As awareness of the economic cost and highly contagious nature of FMD rose, many participants in the meat and livestock trade began to back the introduction of more strict, wide-ranging control measures.[60] However, the resulting restrictions fundamentally altered their experiences of FMD and brought it to the notice of a far larger section of society than ever before. For example, the appearance of FMD led to the drawing of infected areas within which livestock movements were prohibited except under licence. This meant that FMD became a geographical problem situated within a particular region, rather than a biological problem located within the animal body. Whereas previously FMD had mattered only to the owners of diseased stock, it now affected the daily lives of all participants in the meat and livestock trade within the surrounding district (see Plate 5). During the early 1880s, the government began to cancel fairs and markets in FMD-infected areas. This not only inflicted severe financial losses; it also enhanced feelings of isolation among rural communities, for whom the weekly market day was the social highlight of the week. As the hardships inflicted by FMD control grew, livestock owners began to fear the appearance of disease and threw themselves into the campaign for import restriction in the hope that this would halt its invasion and remove the need for irksome, costly domestic trade controls.[61] At the same time, Parliamentary opposition to import restriction declined, partly out of sympathy for the farming plight, but also because the recent expansion in the dead meat trade meant that many Liberal MPs no longer feared for the meat supply. These developments meant that a complete ban upon live-stock imports – a measure unthinkable just a few years previously – became more generally acceptable and passed into law in 1884.[62]

The 1884 Act had the desired effect as FMD ceased to spread and the nation became free of infection for the first time since 1839. It settled a long-running feud by proving the significance of FMD importation and the need for restrictions upon the foreign livestock import trade. It also brought about further changes in popular experiences of, and attitudes towards, FMD. Previously, FMD had been a familiar everyday event; but as the months stretched into years with only sporadic, easily extinguished, outbreaks, memories faded and it became an unknown and alien invader that appeared without warning and spread in mysterious ways. And because national freedom from FMD was now the norm, even a single case of disease was more visible than ever before, provoking widespread comment and alarm.[63]

Meanwhile, other FMD-free countries such as the US, Canada and Australia decided that they, too, should ban livestock imports from FMD-infected nations. Although Britain was a net livestock importer, a handful of politically influential pedigree breeders regularly sold valuable animals to overseas buyers.[64] They were extremely upset by the export trade restrictions that followed the reappearance of FMD in Britain, and began to press the veterinary department to slaughter infected livestock in the belief that this would eliminate FMD faster than isolation. Brown initially resisted their lobbying. He claimed that slaughter was too costly, and that most livestock owners and local authorities preferred isolation, which worked well. He was prepared to permit local authorities to slaughter if they wished; but few did so, partly because they were required to pay compensation, but also because they were unable to prevent disease reinvading from neighbouring counties. However, as time went on, Brown's resistance weakened, and when FMD reappeared in 1892 he resorted to slaughter to defeat a stubborn pocket of infection. Later that year, a new Diseases of Animals Act enabled the Privy Council to order the slaughter of, and provide compensation for, FMD-infected animals.[65] In the sporadic outbreaks of subsequent years, Brown's successors turned increasingly to slaughter. Initially, they used it only in the first or last few cases of each outbreak in an attempt to speed up disease elimination. But as Chapters 2 and 3 will reveal, it eventually became a universal and compulsory policy.

So, by the turn of the 20th century, a new vision of FMD had taken hold in Britain under the influence of legislation that would endure for over a century. Memories of the earlier controversy had faded, and most people found it incredible that anyone could ever have believed FMD an inconsequential ailment and resisted its control by the state. Yet, arguably, if the cattle plague had not invaded Britain, FMD would never have become a plague. It would have continued to evade public and political scrutiny; legislative FMD controls would not have been introduced; the slaughter of infected animals would have remained an unthinkable act; and FMD would have continued to be a private farming affair instead of a high-profile public issue with major social and economic implications for entire rural communities. Of course, this is all speculation; but the point is that FMD-as-plague was a 'manufactured' problem, the emergence of which was inextricably linked to the social, economic and political concerns of mid to late Victorian Britain.

The consensus generated during the late 19th century upon the nature and control of FMD did not last long. It was merely the first in a succession of temporary lulls that occurred whenever FMD was absent from Britain and legislation appeared to be working. Before long, FMD would reinvade and spread inexorably, and disputes about its nature and control would begin all over again. Next time, however, few would dispute the fact that FMD was a dangerous foreign plague that had to be eliminated from Britain.

Chapter 2

The Politics of Plague: Home Rule for Ireland, 1912–1923

THE IRISH QUESTION

The years 1912–1919 saw rioting, bloodshed and finally civil war in Ireland as rival factions fought over its political future. Since the 1801 Act of Union, Ireland had been governed from London and sent 100 MPs to Westminster. But the later 19th century saw an increasing demand for national self-government. William Gladstone, then leader of the Liberal party, twice tried to repeal the Act of Union and allow Ireland a parliament for domestic affairs. His first attempt took place in 1886, when 93 Liberal MPs voted against the government and defected to the Tory or Unionist party, forcing his resignation. On returning to power in 1892, he tried again. This time the bill passed the Commons, only to be thrown out by the House of Lords. But the issue did not go away, and when, in 1910, two general elections returned hung parliaments, John Redmond, the leader of the Irish party, seized his chance. He offered to support Liberal leader Herbert Asquith in return for Home Rule.

Redmond's backing enabled Asquith to force through the 1911 Parliament Act, an important piece of legislation that prevented the House of Lords from permanently vetoing legislation passed by the Commons. Its passage meant that the third Irish Home Rule Bill, introduced in 1912, was more likely to succeed than its predecessors. Nevertheless, Irish Home Rule remained an incredibly contentious issue, with Parliament and the nation divided over whether Ireland was fit for self-government, what form this should take, and what should be done about Protestant Ulster. As a general rule, Liberals and the nationalist Irish party supported Home Rule, while Ulster Unionists and Conservative politicians opposed it; but the format of the third Home Rule Bill gave rise to misgivings on both sides of the House. After numerous amendments, it passed in 1914; but with the world degenerating into war, its operation was temporarily suspended.[1]

While these events were unfolding, Ireland faced another crisis as foot and mouth disease (FMD) returned with a savage vengeance after long years of absence. Its appearance threatened the mainstay of the Irish economy: the meat and livestock trade. Each year, Irish farmers sent hundreds of thousands of animals across the Irish Sea to industrial Britain, where demand for meat considerably outweighed the mainland supply. Making the journey were fat stock for slaughter, and store animals for further fattening in the rich grazing lands of Scotland and East Anglia. Farmers, dealers, middlemen, importers, ship-owners and butchers all participated in the trade, and animals changed hands many times before arriving at their final destinations.[2] On several occasions between 1912 and 1923, the veterinary inspectors of the British Board of Agriculture diagnosed FMD in animals recently landed from Ireland, and controversy erupted as parties on both sides of the Irish Sea accused each other of harbouring infection. At the heart of these disputes lay several contentious questions. Was there incontrovertible proof that the disease in question was FMD? Had animals contracted infection before or after their departure from Ireland? Were Irish farmers and their Department of Agriculture and Technical Instruction capable of identifying and responsibly managing FMD? And what measures were justified in the name of FMD control? Therefore, running alongside the battle for Irish Home Rule was another battle over the origin, spread and control of FMD. As we shall see, the former powerfully influenced the latter, so that, once again, FMD became a highly politicized affair.

In the spring of 1912, as the House of Commons debated the third Irish Home Rule Bill, a Liverpool offal dealer announced that he had found FMD-like blisters in the tongues of slaughtered cattle that had recently been imported from Ireland. This news was conveyed to Stewart Stockman, the chief veterinary officer (CVO) of the Board of Agriculture, who confirmed that the animals were, indeed, infected.[3] This was alarming news. Both British and Irish agricultural authorities had thought the nation free of FMD and were confident that the controls on livestock imports would prevent new disease invasions. Now, however, it seemed that an undetected focus of FMD existed somewhere in Ireland, and had managed to evade the attention of livestock inspectors stationed on both sides of the Irish Sea.

As we have already seen, 19th-century debates upon FMD led to its recognition as a highly contagious ailment that caused costly reductions in meat and milk production. Consequently, on its return in 1912, there was no doubt in anyone's mind that it had to be eliminated. T W Russell, an

Ulster Presbyterian and Unionist who headed the Irish Department for Agriculture and Technical Instruction, immediately directed his veterinary inspectors to discover the whereabouts of FMD. They traced the infection to a farm near Swords in County Dublin, where an unqualified cattle doctor was in attendance. The cowman was prosecuted for failing to notify the authorities of the presence of FMD, and all remaining livestock on the farm were slaughtered. Then began the painstaking task of tracing all of the animals that had come into contact with the diseased during their journey from Swords to Liverpool. On discovering an infected farm, inspectors slaughtered clinically sick animals and isolated the rest. They also halted livestock movements, fairs and markets in the surrounding infected area, and called in the Royal Irish Constabulary to enforce these restrictions (see Plate 6).

Stockman and the veterinary inspectors of the Board of Agriculture were similarly occupied on the mainland. However, unlike Russell, Stockman chose not to apply the discretionary policy of slaughter that he had inherited from his predecessors and, instead, ordered the slaughter of all livestock upon infected farms. This, he argued, was a more effective measure as, normally, FMD symptoms did not appear until several days after infection had occurred, so that animals which appeared healthy might actually be spreading the disease. His policy proved popular with leading pedigree breeders because in speeding up the elimination of FMD, it promised to allow the rapid resumption of exports to FMD-free nations.

Stockman soon discovered that FMD had spread to a number of different mainland locations. On receipt of this alarming news, the board's president, Walter Runciman MP, decided to stop all imports of Irish livestock until Russell's staff had discovered the full extent of FMD in Ireland.[4] This news dealt a catastrophic blow to participants in the Irish meat and livestock trade, and was strongly denounced by the Irish party MPs John Dillon, T M Healy, James Farrell and William O'Brien, who proclaimed it an unnecessary and unjustified measure.[5] However, British upper-class landowners and Unionist MPs, such as Henry Chaplin (a past president of the Board of Agriculture who had led the 1880s drive for livestock import prohibition), Charles Bathurst and Captain Pretyman, welcomed the news on the grounds that Britain had to be protected from further invasions of a disease that endangered both agriculture and the pedigree livestock export trade.[6]

The two sides battled over FMD control in Parliament and the press. Soon it became clear that they were not fighting simply over how best to defeat a terrible animal plague, or over how to minimize the financial costs of FMD control. Although these were important factors, a far more weighty concern was the implication of FMD for Home Rule. Each side drew connections between Ireland's capacity to understand and control FMD, and its ability to understand the principles of self-government and control its own people. Consequently, debates became extremely emotive and achieved a high political profile.

As far as Chaplin, Bathurst, Pretyman and their colleagues were concerned, Irish reactions to FMD were totally inadequate. Chaplin had held this view since the 1870s and 1880s, when Irish farmers and dealers had resisted legislative controls upon the movement of infected animals and claimed that FMD was not a serious problem.[7] Consequently, he was not surprised to hear rumours of Irishmen evading livestock movement restrictions and smuggling cattle out of infected areas at night. He argued that in the absence of the trade embargo, the Irish would spread FMD willy-nilly throughout the mainland. Bathurst focused his attack upon the Irish Department of Agriculture, claiming: 'Ireland owes her present misfortunes and England the recent serious recrudescence of this disease to the ignorance, weakness and dilatoriness of the Department which Mr T W Russell now controls.'[8] He argued that Russell and his team had probably known all along that FMD was present in Ireland – after all, the Irish were so talkative that they were incapable of keeping a secret! Certainly, it should not have taken Russell three days to trace the disease to Swords, and he was completely mistaken in culling only visibly infected animals and isolating their contacts. In Bathurst's opinion, Russell should have followed the example of the British Board of Agriculture, which had learned, during the sporadic outbreaks of the past 28 years (a period in which Ireland had been free of FMD), to diagnose and control the disease properly. He even argued for the transfer of responsibility for contagious animal disease control from the Irish department to the Board of Agriculture, a move that would have simultaneously undermined the Irish fight for Home Rule.[9]

Unsurprisingly, such derogatory comments did not go down well in Ireland. Healy and his colleagues refuted the charges made against Russell, and begged Stockman and his staff to come and witness for themselves how effectively Irish inspectors could control FMD. Stockman refused; therefore, to prove its competence, the Irish Agricultural Department took the extremely unusual step of publishing photographs of veterinary inspectors at work in its annual report (see Plates 7 and 8).[10]

Irish spokesmen argued that a trade embargo was completely unnecessary as FMD only existed in one or two of the seven Irish counties. Nor were they satisfied when, three weeks after the discovery of FMD, Runciman decided to admit small numbers of Irish fat stock providing that they were sent from disease-free areas of Ireland to specified British ports for slaughter on disembarkation. Normally, the fat-stock trade made up only 25 per cent of Irish exports. Much more important was the trade in store animals, which earned UK£10 million each year. Most Irish farmers had insufficient fodder to feed stores during the autumn and winter and relied upon their sale to pay rents or make periodic repayments on their farm purchase loans. Consequently, they continued to agitate for the complete lifting of all trade restrictions.[11]

British agriculturalists reasoned that the board's embargo was only equivalent to that which the Irish department had placed on the British export trade the previous year, when FMD had broken out in Surrey. However, the British export trade to Ireland comprised only a few pedigree beasts worth in total less than UK£100,000 a year. According to the Irish, their trade should receive more lenient treatment because it was the more valuable. On the contrary, said Chaplin. Because the Irish trade was more extensive and involved the crowding together of hundreds of animals in closely confined conditions, it was far more likely to spread disease than the British export trade, which transported animals singly, under close supervision. Instead of complaining about the situation, Irish farmers should show some initiative and adapt to the changing conditions of trade. For example, they could learn how to grow winter fodder, or construct abattoirs so that animals could be killed before export to the mainland.[12]

Unionist breeders claimed that Irish farmers had obviously failed to understand the severity of the situation, otherwise they would not have lobbied for the resumption of trade at a time when fresh FMD outbreaks appearing in Ireland. They also alleged that farmers on the mainland had grown mistrustful of the Irish animals, a situation that could only be rectified if Irish farmers quietly accepted the trade embargo.[13] *The Times*'s agricultural correspondent strongly supported such views, while claiming that the Irish reaction to the trade embargo 'shows a lack of experience in matters of the kind – and, it has been suggested, her dislike of legislative control'. Following this logic, Irish farmers were damned if they complained about the embargo and damned if they did not. However, not all British interests agreed with Chaplin and his colleagues. Graziers from Norfolk and Scotland, who usually bought up Irish stores for fattening, demanded the swift resumption of trade, as did shippers, auctioneers, dealers, salesmen, butchers and the National Federation of Meat Traders.[14] Breeders stood firm, however, dismissing such representations as 'selfish', while claiming that the embargo was 'in the interests of the nation'.

One prominent Irish criticism was that in its response to FMD, the Board of Agriculture had treated Ireland as it would a foreign infected country. Spokesmen contrasted the board's total ban upon all Irish exports with the 15-mile trade exclusion zone that it drew around infected farms on the mainland. They also pointed out that the 19th-century legislation under which the embargo had been imposed was only ever intended for use against foreign countries. The board dismissed such complaints, saying that it normally banned exports from foreign FMD-infected nations for a period of six months, while the Irish trade was resumed, in part, after only three weeks. This statement failed to quell Irish suspicions because, on readmitting Irish fat-stock exports, the board required their landing at 'foreign animals' wharves', isolated areas of mainland ports that had been built during the 1870s to receive animals from cattle plague-infected nations. Healy later complained to Parliament that 'the doctrine apparently is that we are to have all the disadvantages of the Union and none of its advantages'.[15] However, since this argument was open to the interpretation that Ireland had no desire to be categorized as a 'foreign' nation, Irish nationalists had to reign in their complaints in order to advance their fight for Home Rule.[16]

As the dispute continued, Unionists and Irish nationalists accused each other of attempting to make FMD control a political issue, while denying that they personally viewed it as such.[17] Irish party MP John Dillon described the trade embargo as 'a most infamous and scandalous conspiracy, largely political and supported by powerful interests in England to maintain unreasonable and unnecessary regulations against the Irish cattle trade'.[18] One of his colleagues, J P Farrell, claimed that English supporters of the embargo were actually trying to ruin the Irish economy,[19] while Healy argued that the British government had a moral responsibility to protect the Irish livestock trade.[20]

The summer passed and the trade still did not return to normal. Irish party members complained to Parliament that their people felt 'persecuted and coerced' and that the whole affair was a Unionist conspiracy.[21] Healy declared: 'This is imperialism – that the Ministry of the Irish Department is not even consulted, much less trusted, by the Minister of the English Department, who is supposed to have the same interests at heart.'[22] Irish grievances were heightened by the case of the 'Waterford head', a pig's head of suspected Irish origin that was discovered bearing FMD lesions in a Liverpool market in July 1912. According to one of the board's veterinary inspectors, the tongue had been deliberately removed in order to conceal the presence of infection, a view later repeated in Parliament as proof of Irish subterfuge. In fact, subsequent investigations showed that the removal of the tongue was a normal practice in the dead meat trade, and that there was no evidence to suggest that the head came from Waterford or, indeed, from anywhere else in Ireland.[23]

Runciman's decision to readmit small numbers of Irish fat stock infuriated Bathurst and his Unionist colleagues, who warned repeatedly of the dire consequences that would follow any relaxation of the trade embargo. Meanwhile, encouraged by a campaign in the *Irish Times*, Irish agitation against the trade restrictions grew. Graziers asked why John Redmond had not secured the lifting of the embargo when only recently he had boasted that he held the British government in the palm of his hand.[24] Farrell and William Field, president of the Irish Cattle Traders' and Stockowners' Association, felt that in protest at the board's handling of FMD, Redmond should withdraw Irish party support from the Liberal government, thereby forcing a general election. However, Redmond knew full well that if he brought down the government, the Liberals might not regain power, and that if a Unionist government was returned, it would not proceed further with Home Rule. So, for the sake of national self-government, he refused to take drastic action against the embargo.[25] This was a wise move, as bringing down the Liberal government was a prime goal of leading Unionists. They sought an end to both Irish Home Rule and Lloyd George's 1909 'people's budget', which had increased the taxes and death duties of upper-class landowners.[26]

Russell, an Ulster Unionist, and Runciman, a Liberal, were somewhat isolated from the increasingly polarized debate between Irish nationalists and English Unionists, and their attempts to reach a compromise over FMD control did not please either side. During August and September 1912, Runciman gradually relaxed restrictions upon Irish fat-stock exports. In early October, as Irish FMD incidence fell further, he introduced a new Animals (Landing from Ireland) Order, which permitted limited imports of store animals, subject to their inspection and quarantine at the ports and detention at the farm of destination.[27] Chaplin was incensed, and tried to persuade the House of Commons to reconsider the order. He alleged that Runciman was inexperienced in FMD control and had yielded, wrongly, both to Irish pressure and to the Liberal party's self-interested desire to stay in power. He grew even angrier several days later upon learning that several new cases of FMD had appeared in formerly disease-free areas of Ireland. When Runciman refused to amend his plans, Bathurst turned to the local authorities, who at that time were able to regulate animal movements within their boundaries. At his behest, all but six forbade the entry and movement of Irish stock.[28] The Irish were also unhappy. The additional layers of bureaucracy that the new order had imposed upon the trade impeded its smooth running and reduced profits, and they were furious when, in January 1913, the government made it a permanent measure. By then, however, FMD had disappeared from Ireland and England after causing 68 and 83 outbreaks, respectively. As previously infected areas were released from restrictions, exports to the mainland began to increase and debates upon the control of FMD temporarily died away.[29]

For many Irish and English politicians and livestock owners, this episode affirmed their existing attitudes towards Irish Home Rule. As far as Chaplin, Bathurst and their colleagues were concerned, Ireland's reaction to FMD epitomized all of the reasons why it was unfit for self-government. Its people disliked and rebelled against regulations; they were emotional, irrational and irresponsible; and they failed to place the interests of the nation ahead of individual desires. But for Irish nationalists, the board's activities and the comments of leading agriculturalists reinforced their deep-seated conviction that only Home Rule could achieve justice for Ireland. They saw the trade embargo as a punitive measure, an act of imperialism and protectionism that was proof of their unfair treatment at the hands of the English.

GETTING TO GRIPS WITH FMD

The disappearance of FMD and the passage of the Home Rule Bill brought about a temporary lull in the controversy; but before long, both items were back on the political agenda. On several occasions between 1913 and 1923, FMD-infected Irish livestock were again discovered on the mainland. Each time the Board of Agriculture placed an immediate embargo on the Irish trade, and each time the Irish objected. Sometimes the embargo was modified a few days later, when the board's veterinary inspectors discovered that their diagnosis of FMD had been mistaken or that infection had not originated in Ireland; on other occasions it remained in place for several weeks. In either case, the Irish reaction was the same, as farmers, livestock traders and politicians protested furiously about the 'unnecessary' disruption to trade.[30] Healy told Parliament: 'It is wished to prove that at the present time all this disease came from Ireland. Irishmen are supposed to have a double dose of original sin. . .and apparently the same thing applies to Irish cattle.'[31]

Meanwhile, the political situation deteriorated as a result of the 1916 Easter Uprising, in which nationalist rebels tried – and failed – to seize power. In its aftermath, Irish voters deserted Redmond's Irish party, which favoured a constitutional solution to the crisis, in favour of the revolutionary movement Sinn Fein. After winning 73 seats in the 1918 general election, Sinn Fein set up an Irish Parliament, the Dáil, and declared an independent Ireland. Around this time, the Irish Republican Army (IRA) began its campaign against any form of British government in Ireland, and as violence escalated, the British government sent its notorious 'black and tan' soldiers to sort out the situation. In an attempt to resolve the situation, Nationalist and Unionist leaders signed an Anglo–Irish Treaty in 1921.

This allowed the 26 southern counties to become an independent Irish Free State, while the six northern counties remained part of the UK. It was not accepted by all Nationalists, however, and civil war continued in the south until 1923.[32]

The 1912 FMD controversy had taught Irish politicians, agriculture officials and livestock owners of the difficulties involved in resisting the Board of Agriculture's repeated trade embargoes. An out-and-out fight against the British government would have jeopardized the drive for Home Rule, while tit-for-tat restrictions on exports from the mainland had only a limited effect because of the small size of the British pedigree export trade. Consequently, from 1913, they adopted a somewhat different tactic and began to attack the scientific basis of the board's actions. In criticizing its inspectors' diagnoses of FMD and tracings of disease spread, they reopened old debates about the nature and behaviour of FMD, and added fuel to the controversy surrounding its control.

Earlier research, undertaken at the close of the 19th century by German scientists Friedrich Loeffler and Paul Frosch, had revealed that fluid-filled blisters in the mouths and on the feet of FMD-infected animals contained the agent responsible for disease. The behaviour of this agent differed from that of other known germs: it could pass through filters that normally retained bacteria, could not be viewed through a microscope and proved impossible to culture. Loeffler termed it a 'filterable virus' and assumed that, in all other respects, it was akin to bacteria. However, because it could not be isolated or visualized, practically the only way of confirming its presence was to take tissue samples from mouth and feet lesions, inoculate them into susceptible livestock and wait to see whether similar symptoms developed.[33] This technique was too complicated and expensive for widespread use, and so, in the field, veterinarians continued to diagnose FMD on the basis of its clinical symptoms and propensity to spread. But matters were not always clear cut. What if an animal showed symptoms of FMD but failed to spread the disease? Or what if it showed symptoms of a highly contagious ailment that bore only a passing resemblance to FMD? It was similarly difficult to decide upon the origin of the infection. Its incubation period was somewhat variable, and, given the possibilities for indirect virus spread, it was usually impossible to say how, where and when livestock had contracted the disease. These ambiguities allowed politicians, veterinarians and participants in the meat and livestock trade to interpret the available evidence in a manner that accorded both with their personal experiences of FMD and with their personal, professional and national interests.

Veterinary inspectors of the British Board of Agriculture tended to diagnose FMD even when the animals concerned showed extremely obscure symptoms and had failed to spread infection to their susceptible contacts. Often, they decided that infection had originated in Ireland, even in cases where animals had left disease-free Irish farms, failed to convey infection to livestock with which they had mixed during the journey, and developed symptoms several days after arriving on the mainland. They explained this occurrence by reference to the indirect spread of virus, and precautionary action against the Irish trade soon followed. By contrast, Irish spokesmen rejected diagnosis on suspicion and required classic foot and mouth blisters before they would admit to the presence of FMD. They also argued that the board's tracing of disease spread was lacking in evidence and could not justify a trade embargo.[34]

History was partly to blame for these differences of opinion. As we have already seen, rigorous restrictions upon livestock imports caused FMD to disappear from Ireland and the mainland during the 1880s. It seemed that the source of FMD – foreign infected livestock – had at last been detected and defeated. The resulting absence of FMD in Ireland meant that this view held sway until the 1910s; but, in the meantime, mainland opinion had moved on, mainly as a result of sporadic reinvasions of FMD that could only be explained by reference to indirect contagion. In 1908, the board restricted imports of hay and straw after they were implicated in an Edinburgh FMD outbreak, and with the appearance of six unconnected outbreaks in 1911, Runciman asked a departmental committee to consider further measures against the disease. Numerous witnesses told the committee of their suspicions that FMD could enter Britain in infected milk, hides, vaccine lymph, offal, vehicles, vegetables, straw and many other substances. Its subsequent report helped to raise awareness of the many different ways in which the invisible FMD virus could enter Britain.[35]

New ambiguities about the clinical picture of FMD came to light following the 1912 Swords outbreak, when the Board of Agriculture decided that all cattle leaving Ireland should be 'mouthed' – examined to see if they showed FMD-like oral lesions (see Plate 6). This practice led to the discovery of a huge variety of pathological changes. Previously, veterinarians and farmers had not been in the habit of looking inside cows' mouths, and so substantial doubt arose over what was, and what was not, characteristic of FMD. Inspectors also tried to age the lesions they discovered – on that basis, they could decide when (and, more importantly, where) animals had contracted disease. However, this was a rather inexact art and inspectors' conclusions were open to criticism.[36]

Compounding these difficulties were the divergent interests and preconceptions of veterinary surgeons, farmers and politicians on either side of the Irish Sea. The circumstances surrounding the 1912 Swords outbreak had aroused mainland suspicions that FMD lurked, unnoticed, upon Irish farms, while the popular stereotype of the Irishman as a talkative, unreliable and ignorant individual meant that Irish denials were not taken seriously.[37] As far as veterinary inspectors of the board were concerned, if a disease resembling FMD appeared in recently imported Irish stock, then it was best to play safe and confirm the diagnosis because a furore would erupt should they fail to prevent FMD from spreading. (Interestingly, they were not so strict about FMD in British livestock, preferring to 'wait and see' when diagnosis was in doubt.) For the political reasons already discussed, aristocratic landowning Unionist MPs supported their actions, as did many livestock owners, who in the absence of Irish competition could obtain higher prices for their own animals. The Irish, on the other hand, were keen to clear the clouds of suspicion that surrounded the health of their cattle, and to avoid the profound economic and political consequences of trade embargoes. Consequently, they demanded substantial, incontrovertible evidence before they would admit that a disease was FMD and had originated in Ireland.[38]

The many ambiguities surrounding FMD meant that neither side could win this scientific dispute; as a result, the board continued to halt the Irish trade on suspicion of FMD and the Irish continued to criticize its actions. However, in 1916, following repeated Irish jibes about its mistaken FMD diagnoses and implausible disease tracings, the board's new president, Rowland Prothero, decided that the two sides should try to clear up the dispute. He suggested a joint investigation, 'by means of experiment or otherwise', into methods of diagnosing FMD. T P Gill, secretary of the Irish department, replied that he was happy to cooperate in investigations if they helped to remove the uncertainty felt by the board's veterinary officers, but that his veterinarians required no such aid. Stockman retorted that if the department did not participate, he would continue to 'play safe' and diagnose FMD on suspicion. This forced the department's hand, and in 1917, a Joint Committee on Stomatitis was appointed.

Committee members included Professor John McFadyean, Stockman's father-in-law and RVC principal, and the head of the Dublin Veterinary College, Professor Mettam, who was soon to die of pernicious anaemia. They tried to culture the disease agent and to reproduce symptoms by inoculating material taken from disease lesions into susceptible animals. Results were distinctly unremarkable. After taking evidence from practising veterinarians on how best to diagnose FMD, the committee came up with the hardly earth-shattering conclusion that it was often difficult to reach a definite diagnosis using mouth lesions alone, and that other characteristics, such as fever and the ability of the disease to spread, should be taken into consideration. The report was not published on the grounds of insufficient public interest.[39]

In 1920 the Irish trade was again halted, and then resumed when British veterinary officials decided that the disease in question was not FMD. Similar events occurred the following year when FMD broke out in Derbyshire among cattle recently imported from Ireland. The Irish department protested in vain that it had suffered only one outbreak of FMD since 1914, and several months later, when FMD really was confirmed in Ireland, Stockman made plain his belief that it had existed all along.[40] In 1922, a devastating FMD epidemic took hold on the mainland, and English farmers' suspicions again fell upon Irish livestock imports. In this case, Stockman claimed that there was no evidence that FMD existed in Ireland. However, he thought farmers' complaints perfectly understandable given the Irish habit of denying FMD in the face of overwhelming evidence. He went on to launch a scathing attack upon ignorant Irish dealers, who had helped to propagate FMD by shifting livestock over large distances and through numerous fairs. His veterinary inspectors had no jurisdiction over these men, who had usually returned to Ireland by the time their involvement was discovered. They further hindered the board's attempts to trace FMD spread by failing to keep records of livestock sales, often because they could not read or write.[41] The departmental committee that later enquired into the 1922 epidemic also took a dim view of Irish dealers' habits, although this was hardly surprising given the identity of its chairman – Captain Pretyman, Unionist MP and arch-critic of the Irish trade.[42]

The British tendency to scapegoat the Irish diminished markedly after 1923. Perhaps this was because FMD disappeared from Ireland, although the absence of disease had not prevented past controversies from erupting. Or possibly it resulted from new legislative restrictions upon Irish dealers' activities, introduced on Pretyman's recommendation, to prevent them from moving animals rapidly between markets and fairs.[43] However, it was no coincidence that controversy over FMD ceased at about the same time as debates over Irish Home Rule. With the settling of the Irish question in 1923, relations between the two countries substantially improved. It was then no longer necessary for Irish and mainland interests to use FMD as a tool with which to express their opinions of each other and to further their economic and political agendas. But as Chapter 4 shows although FMD became de-politicized within the context of Anglo–Irish relations, it was soon to restructure Britain's political and economic dealings with a very different country, Argentina.

Chapter 3

The Epidemics of 1922–1924

OVERVIEW

Between 1922 and 1924, rural Britain experienced two of the worst foot and mouth disease (FMD) epidemics in 40 years. Their unknown origin and rapacious spread caused anxiety and panic to livestock owners, and for months on end, the veterinary staff of the Ministry of Agriculture and Fisheries (MAF, which took over from the Board of Agriculture in 1919) struggled desperately to bring FMD under control. Their efforts eventually succeeded, but at a price. FMD control cost the taxpayer millions of pounds and practically halted rural life in parts of Britain. At the height of the epidemics, suffering farmers rebelled, issuing increasingly desperate but ultimately unsuccessful demands that MAF alter its FMD control policy. This chapter relates the events and experiences of those years, focusing upon the plight of Cheshire farmers, who were the hardest hit by the disease.

The origins of the epidemics were never discovered; but in circumstances reminiscent of the 2001 epidemic, the extensive movement of livestock (especially store stock imported from Ireland) was held responsible for the rapid extension of disease throughout the nation. Investigations suggested that FMD-infected animals first appeared in markets in January 1922, during one of the busiest trading periods of the year. They spread infection both directly, to the animals they mixed with, and indirectly, as the market pens, loading bays and railway trucks in which they were held, as well as the clothes and boots of their handlers, became contaminated with the FMD virus. MAF was unaware of the presence of FMD until alerted by Belgian vets at Antwerp, who discovered symptoms in cattle recently exported from Hull for slaughter. By then the plague had been spreading unnoticed for at least a week, and the seeds of 700 to 800 outbreaks were already sown.

Responsibility for controlling the disease lay with Stewart Stockman, chief veterinary officer (CVO) of the Ministry of Agriculture (see Plate 10). As we have already seen, Stockman believed that slaughtering diseased animals and their contacts was the best method of eliminating FMD from Britain. He argued that unlike the former policy of isolation, slaughter stopped infected animals from manufacturing virus and therefore controlled the disease faster. This, in turn, lessened the need for restrictions upon the movement and marketing of animals in the surrounding area – an important benefit considering the recent increase in the scale and frequency of livestock movements. Slaughter also minimized the impact of FMD upon the livestock export trade, which was subject to restrictions imposed by FMD-free nations such as Canada, the US, New Zealand, Australia and South Africa. Stockman attached great importance to the preservation of this trade, as although economically insignificant, it was dominated by influential upper-class landowners who viewed the foreign demand for their pedigree stock as a sign of British national superiority.[1]

Stockman's penchant for slaughter did not extend to valuable pedigree animals, which he isolated and allowed to recover from FMD. He argued that it was in the national interest to preserve these irreplaceable products of generations of livestock breeding. Another motivating factor was cost: the 65 infected farms that he subjected to isolation during the 1922–1923 epidemic would, if slaughtered out, have increased the compensation bill by 25 per cent. Stockman also believed that in contrast to most farmers, pedigree livestock breeders were educated, influential and responsible individuals who possessed the facilities and the resources to nurse animals back to health, and could be trusted to take adequate precautions against the spread of disease. Therefore, while nominally serving the interests of British agriculture as a whole, Stockman's slaughter policy actually favoured upper-class livestock breeders, who encouraged the slaughter of 'ordinary' stock in order that their animals could escape infection and embark upon lucrative journeys overseas.

The 1922 epidemic lasted eight months. During that time, the short-staffed, poorly equipped State Veterinary Department, numbering just 59 inspectors at the start of the epidemic, struggled to control a total of 1140 FMD outbreaks. Each veterinary inspector took charge of all outbreaks occurring within a set geographical area. They confirmed the diagnosis of FMD, oversaw the valuation and slaughter of diseased animals and their contacts, directed butchers to salvage the meat from healthy carcasses, and organized a supply of manpower and fuel for the burning or burial of diseased carcasses. They also mapped out infected areas, usually 15 miles in radius from the infected farm, within which hunting was banned and livestock movement restrictions of varying severity imposed for several weeks.

It fell to the local authorities to enforce these measures and, if they thought fit, to impose additional restrictions upon livestock movements within their boundaries. They acted through Diseases of Animals Committees, which in turn, appointed inspectors (usually policemen, although by law they had to appoint at least one veterinary surgeon) to execute disease controls. Lay inspectors issued licenses for fat-stock markets and necessary livestock movements, organized the disinfection of railway trucks and pens, and arranged for policemen to guard footpaths and the entrances of infected farms (see Plate 11). Veterinary inspectors examined livestock in markets for signs of infection; they were first on the scene when farmers reported the appearance of disease and had powers to halt livestock movements in the vicinity. In 1922, the disease had a head start, and as notifications flooded in, MAF and local authority inspectors alike were unavoidably delayed in executing the control policy. Eventually, they succeeded in halting disease spread, but not until an unprecedented 56,000 livestock had been slaughtered. Compensation for this loss cost the taxpayer UK£750,000, a vast sum during the financially straitened post-war years.

After 12 months' respite, the disease flared up again. In August 1923, MAF received several simultaneous reports of FMD from widely separated geographic locations. Its FMD control machinery again swung into action; but this time delays in the diagnosis, slaughter and disposal of infected stock were even more marked. The next nine months saw nearly 2691 FMD outbreaks, more than any other epidemic on record, and nearly 300,000 livestock were slaughtered, at a cost of UK£3.3 million in compensation. Cheshire experienced 1385 outbreaks and bore the brunt of the epidemic. For months, FMD ran rampant through the county, inflicting terrible suffering upon its rural inhabitants. The final death toll – 50,000 dairy cattle – amounted to one third of the Cheshire herd, and in the worst affected areas up to 60 per cent of farms were emptied of livestock.

In both epidemics, but particularly during 1923–1924, Stockman's decision to slaughter almost all infected and contact livestock aroused considerable concern. He had used this policy before, during the 1910s; but then, outbreaks had numbered less than 100 a year, and slaughter had provoked little criticism, partly because few experienced it at first hand, but also because it had the desired effect of eliminating FMD rapidly from Britain. In 1922–1924, however, the effects of this policy were felt more widely as thousands of farmers from all over Britain lost their stock and were confined to empty farms, their life's work destroyed. As the weeks passed and disease continued to spread, disappearing only to return upon a far greater scale than before, whisperings against the slaughter policy grew into an immense public outcry. Some of its critics saw no justification in killing animals which, if left to their own devices, would soon recover from FMD; others saw slaughter as a good thing in principle, but became increasingly uneasy at its failure to work in practice. The one question on everyone's lips was whether the legislative 'cure' for FMD was worse than the disease itself.

During the autumn and winter of 1923, the controversy over FMD control reached a level of intensity not experienced since the 1870s. Viewed retrospectively, it amounted to one of the most powerful 20th-century forces for change in the government's handling of FMD. MAF had no experience of slaughter in the control of a raging epidemic, and none could be sure that it would succeed. At the same time, the older isolation policy – which had, during the 1880s, helped to drive endemic FMD from Britain – still carried significant authority. There was everything to play for, and as the crisis deepened, farmers, veterinary surgeons, doctors, participants in the meat and livestock trade, and MAF and local authority officials all pitched into a desperate battle over FMD control.

Stockman and his upper-class agricultural supporters were absolutely convinced that slaughter was the right policy, and that given time it would prove its worth. They employed several key arguments in an attempt to impress their views upon a hostile public. First, they contrasted the immense costs of endemic FMD with the low annual average cost of slaughter. Second, they emphasized that when compared with slaughter, isolation would result in greater trade losses because it required more lengthy restrictions upon livestock movements. Third, they drew favourable comparisons between Britain's FMD incidence and that of Continental nations, which used isolation. Finally, they rewrote history to present slaughter as *the* traditional control policy, which for 30 years had kept the nation relatively free of disease.

As we shall see, Stockman eventually succeeded in controlling both FMD and public opinion, and saw his views on slaughter endorsed by two government-appointed committees of inquiry into the epidemics.[2] His victory was highly significant; having discovered through experience that slaughter could succeed in the face of widespread disease, MAF resolved to use it in the control of all future epidemics, regardless of public opinion. So, paradoxically, the most widespread, authoritative 20th-century challenge to FMD control by slaughter resulted not in the overthrow of this policy, but in its rising authority. During the years that followed, critics of the slaughter made little headway, and this policy was not threatened again until the 1950s, when vaccination appeared on the scene (see Chapter 6).

THE CHESHIRE EXPERIENCE, 1923–1924

*Too late! Too late! will be written of Whitehall by the Historian
of the Cattle Plague in Cheshire in the years of grace 1923–24.*[3]

Cheshire, during the 1920s, was a land of dairy farms populated by
thousands of dairy shorthorn cattle. These 'dual-purpose' cows also
produced good quality beef and were extremely popular until the 1940s,
when they were gradually replaced by the higher-yielding Friesian Hol-
steins. The rigours of hand-milking imposed a natural limit upon herd sizes,
and few farmers owned more than 100 cows. Many bred their own
replacements, selling old cows to the butcher when their milk production
dropped. They were closely bonded to their animals, and often boasted of
their high quality and commercial value. The milk produced supplied the
urban populations of Liverpool and Manchester and was also turned into
Cheshire cheese, a declining practice that, nevertheless, contributed to a
strong sense of local identity. Cheese-making was well suited to the
geography of Cheshire, a flat plain with high rainfall where grass grew all
year round. Instead of housing and feeding their cattle during the winter,
many smallholders stopped milking them and left them outside to graze.
They made cheese in the spring, when milk yields increased after calving,
and again in autumn. This 'low input–low output' system did not produce
particularly large volumes of milk; but it made economic sense.[4]

When, in September 1923, FMD first appeared at Crewe, Cheshire
farmers' prime concern was the circumstances surrounding its entry into
the county. They found MAF officials 'close as oysters' on the subject, and
so J Sadler of the Cheshire Chamber of Agriculture turned local detective.
He unearthed evidence to show that the mistaken actions of two veterinary
surgeons had allowed FMD to take hold. One had failed to disinfect his
clothing after inspecting infected animals at Blackpool; in a subsequent call,
he carried virus to Fleetwood docks, where it infected livestock in transit
from Ireland to Crewe. Shortly afterwards, rumours of the presence of
FMD spread through Crewe market, and there was a rush to move animals
out of the area. The veterinary surgeon in charge of the market, Mr
Manuel, eventually imposed restrictions upon livestock movements; but it
was all too late. Infected animals had already moved deep into Cheshire,
scattering FMD virus in their wake. Manuel later defended his belated
action, explaining: 'You must consider, I have to live by my profession in
the district and if a man certifies and stops the whole countryside he is
practically done in his practice in that district . . . had it turned out at a
subsequent investigation not to have been FMD, people would have
deemed me a fool.' Most Cheshire farmers sympathized with his dilemma,

and reserved their criticisms for those MAF officials who had tried to cover up the root cause of disease spread.[5]

After an uneventful month, Cheshire experienced a sudden explosion of FMD cases late in October. Shortly afterwards, Stockman visited the county and discussed the situation at a public meeting at Whitchurch. This was a restrained affair. Farmers listened politely as he explained that shortages of manpower were preventing his staff from making progress against the disease, which was acting strangely and 'jumping' between farms. He encouraged farmers to take precautions against disease spread by limiting their social contact with other stockowners, placing animals indoors, and disinfecting people and vehicles leaving and entering the farm. He also argued in favour of the current policy, claiming that the alternative, isolation, was more costly and less likely to control the disease.[6]

During the weeks that followed, FMD continued to spread unchecked, and increasing numbers of farmers experienced alarming delays in the valuation, slaughter and disposal of their infected stock. By 24 November, the county had lost 3000 cattle, and as their patience with MAF wore thin, farmers began to demand greater local participation in FMD control. They alleged that veterinary practitioners could diagnose disease and slaughter infected stock far quicker than MAF inspectors, and called on Stockman to delegate tasks to the county council, which was more in tune with local conditions than ministry men.[7] Stockman regarded such demands as a challenge to his personal and professional authority, and rejected them outright. His ongoing refusal to deviate from a highly centralized and seemingly inefficient method of disease control did little to dampen farming criticisms, and provoked quarrels between local and central government officials.[8]

Early in December, the tide of Cheshire opinion began to turn against the slaughter policy. By then, MAF was receiving 60 notifications of FMD a day from Cheshire alone, and the rising death toll and growing delay between diagnosis and slaughter led many advocates of the latter policy to admit that in Cheshire, at least, it had failed. Cheshire Farmers' Union decided to reconsider a resolution it had recently passed in support of slaughter. At a meeting attended by National Farmers' Union (NFU) President Harry German, members discussed the spread of disease and made tentative suggestions of alternative control policies. German spoke out in support of the slaughter. He told farmers that they had to consider the rest of the nation as well as themselves, and that Stockman had no intention of changing his policy. Most of the audience ignored him and voted in favour of a resolution that condemned slaughter as ineffective, and demanded its immediate replacement by 'protective and preventive measures'.[9]

This meeting marked the start of German's efforts to persuade Cheshire farmers to accept the slaughter policy. It seems that he regarded the 1923–1924 FMD epidemic as a prime opportunity to gain political power and influence for the NFU, a body established 15 years before by a group of Lincolnshire tenant farmers. Whereas the Royal Agricultural Society (RAS) and Chamber of Agriculture were upper-class organizations made up of landowners, aristocrats and members of parliament (MPs), the NFU mostly consisted of tenant farmers who lacked parliamentary influence. German aimed to enhance its profile by building direct links with MAF, a strategy that came to full fruition during and after World War II, when MAF regularly consulted the NFU regarding subsidies for agriculture.[10] In conversation with Stockman, he portrayed the NFU as a responsible body made up of educated farmers who were prepared to suffer the hardships of the slaughter policy for the good of the nation. He also attended farmers' meetings where he urged compliance with MAF's disease control policy. At the height of the epidemic, he went so far as to appeal 'to branches of the Union not to send up resolutions to HQ advocating that the government should be asked to stop their policy of slaughtering'. Memos and press releases issued by the NFU council further advertised its backing for slaughter, while advocating farming vigilance and listing the precautions necessary to prevent disease spread.[11] The beleaguered Stockman grew to depend upon the NFU's support, and the Minister of Agriculture rewarded German with appointments to both the 1922 and the much smaller 1924 committee of inquiry into the FMD epidemics.[12] In Cheshire however, German's well-known support for the slaughter policy and proximity to Stockman made him an object of suspicion. Farmers called him an 'apologist for Whitehall' and resented his attempt to sway local debates upon FMD.[13]

MAF's delays in controlling FMD forced 64 Cheshire farmers to wait over 14 days between disease diagnosis and slaughter. During that time, many took veterinary advice upon how to alleviate the symptoms of disease. They cleaned and dressed the feet of infected animals and supplied nourishing feed, and were surprised to discover that after several days, the majority had recovered. Several owners of infected pedigree livestock that had been exempted from the slaughter reported similar experiences, while older farmers and veterinary surgeons recalled the ease with which animals had been cured of FMD during the 1870s and 1880s.[14] Such reports suggested that FMD was not the terrible and devastating plague that Stockman had made out, and encouraged further resistance to the slaughter policy.

A number of farmers tried to conceal FMD from the authorities and nurse their animals back to health. One alleged cure and preventive that gained a high reputation in Cheshire was Dr Shaw's remedy. George Thomasson, a farmer from Nantwich, and Sydney Barker, a Salop farmer and registered veterinary practitioner, purchased several batches after reading an advertisement in the press and administered them on over 20 farms, with reported success.[15] On investigating the matter, Vincent Boyle, a MAF veterinary inspector, found Barker 'entirely ignorant' upon the subject of FMD and reported that many of the animals 'successfully' treated had never actually suffered from FMD. However, the enormous local faith in Dr Shaw's remedy caused the 1924 committee of inquiry to override MAF's objections and commission a scientific test, in the hope that this would convince Cheshire farmers of its worthlessness.[16]

Although many animals had practically recovered by the time MAF veterinary inspectors arrived, they were slaughtered anyway, much to the fury of their owners. The local press publicized the case of Mr Winward, a Malpas farmer who had spent a fortnight nursing his animals back to health. He then argued that because they were no longer contagious there was no need to kill them, and appealed to German to intervene on his behalf. German refused, and soon the slaughter team arrived. They killed Winward's 100 cows, of which 94 were due to calve in spring, and returned several days later to slaughter his pigs. He told a friend: 'This place is hell! They are pole-axing in the shippons and shooting them in the field; 13 are now on the fire; the stench is unbearable, I am going away.'[17]

Reports of MAF cruelty and incompetence abounded, fuelling further demands for a change in FMD control policy.[18] One classic account was delivered by farmer Joseph Willet to the 1924 committee of inquiry in FMD:

> I saw on 17th December 14 heifers in a field adjacent to the county Main Road near Calverly Station. Two at least were visibly affected with the disease and were slavering. I saw them again on the 18th. One animal had its head through the rails on the footpath. The newspaper boy walked before me and swished at it as he passed with his bag of newspapers. On Sunday, 23rd December, the 14 were collected on a railway bridge, penned in by gates, area about 100 square yards. An inspector came from Whitchurch at 10.30 am. He had forgotten his killer. He enquired where he could borrow a gun and secured an ordinary sporting gun from the Davenport Arms. He took it with a pocketful of cartridges, went on the bridge and while the church bells were ringing he blew out the brains of the 14 heifers. A visitor from the inn had followed the inspector and was a witness to the act. This man was summoned for going onto an infected

> *farm without a permit. Just about this time the police officer in the local court stated that it was an offence to destroy, by killing, one animal in the sight of another. Information was laid as to the shooting on the bridge. The police declined to prosecute and withdrew the summons taken out against the witness. The animals were left lying on the bridge until the following Tuesday when a local waggoner brought a number of horses in traces and dragged them to a hole prepared 300 yards away.*[19]

A COUNTY UNDER SIEGE

By mid December, 7000 Cheshire cattle had lost their lives. Local newspapers adopted an increasingly desperate tone and published evocative accounts of the conditions experienced by those under siege from FMD. Gangs of labourers – whom many farmers blamed for spreading the disease – proceeded daily from Nantwich to dig trenches, shoot cattle, build pyres and feed the flames with dead animals (see Plate 14). Lorries bearing coal and wood traversed the countryside. Farmers experienced 'the agony of listening to the Greener killer punctuate the moments when death passes along lines of assembled cattle'. In villages, men cursed while women wept. At night, chains of fires lit up the countryside, the smoke 'defacing it with its sinister envelopment and penetrating to farms far beyond'. Those left unaffected withdrew into their homes to wait out the siege, locking gates and placing disinfectant soaked straw across paths, summoned to the outside world only by bells placed outside the gate. They rounded up their cattle and placed them indoors for safety; but still there was no respite to 'the biggest blow that our local agriculture has suffered in living memory' and 'the horrible pest extends, till ruin stares the farmer in the face'. Footpaths were closed, hunting stopped and social functions were cancelled; all agricultural business ceased, and labourers, who were not then entitled to claim unemployment benefit, were thrown out of work and left destitute.[20]

The smoke and smell from the pyres was particularly pervasive. According to a *Daily Mail* journalist, 'Over a wide area the strong, sickly smell of burning flesh is inescapable and during the daytime, clouds of smoke drift slowly over the fields.'[21] Many believed that the smoke was responsible for disseminating virus, a view given scientific credibility by Dr Sloane, a medical officer at the county sanatorium, who believed that fires did not reach a high enough temperature to kill the virus.[22] The 1924 committee of inquiry later addressed this issue. Members' views are illustrated in the following exchange, which also provides insights into the respective characters of German, a hard-nosed, arrogant man, and Stockman, who was more sympathetic to the rural plight:

> German: *'I went to some burnings during an outbreak; I did not think there was anything offensive about it at all.'*
> Stockman: *'No, there is not; but the people were suffering from nerves; they saw the fires blazing at night and it kept it all before them.'*
> Captain Pretyman (chairman): *'It was an additional horror?'*
> Stockman: *'It was, from the farmers' point of view. . .it is really a sentimental feeling and we are prepared to meet that where it arises. . .you could not look out of a window or go out of doors without seeing a fire and it kept the whole horror of the thing before them all the time.'*
> W Bromley-Davenport: *'It was rather awful in Cheshire?'*
> Stockman: *'Yes.'*
> W Smith: *'To see those continuous fires exercised a very depressing influence on the people generally and contributed very much to the antagonism to the policy of slaughtering.'*
> Stockman: *'I think we should consider that sort of thing if human sentiment is involved, and bury where we can.'*
> German: *'It seems to me very weak-kneed, sir.'*
> Stockman: *'You have had, as we have had, the letters of complaint that came in. It must be very real in the minds of these people.'*
> German: *'Their minds are not very strong?'*
> Stockman: *'Do you think the majority of people's minds are very strong?'*
> Bromley-Davenport: *'There is a very widespread belief that the fires did spread the disease, however wrong it may be from a scientific point of view; but from a practical point of view these people did believe that the smoke was, as they said, spreading the disease.'*[23]

As FMD continued to rage, farmers grew anxious that soon there would be no dairy cows left in Cheshire, and called for the preservation of the few stock that remained, without which all hope of regaining the traditional way of life would be lost. They railed against MAF's policy of restricting the 'privilege' of isolation to men of influence, and demanded its extension to their young and in-calf dairy cows. But Cheshire was a lone voice, which failed to dissuade national agricultural organizations from passing repeated resolutions in support of the slaughter policy. In desperation, the *Crewe Chronicle* cried: 'IT IS THE VOICE OF THE COUNTY WE WANT, not that of the country, the NFU or any other organization.' It urged farmers to take matters into their own hands, for the devastation was too terrible to wait any longer.[24]

As the situation in Cheshire intensified, criticisms of incompetence, vacillation, disorganization and ignorance were heaped upon the veterinary department, and Stockman was forced to defend his chosen policy in memos to the press and in written responses to the letters that poured into his office. He later complained: 'There was a newspaper stunt organized against us, and I had to come off what I call my natural work and do a great deal of this sort of fighting to stiffen up the position. . . I should never have been asked to do it.'[25] The 1922 epidemic had already taken its toll upon Stockman's strength, and he had complained bitterly to the first committee of inquiry of the appalling working conditions that he and his staff had faced:

> *People in charge like Mr Smart, Mr Piggott, Mr Jackson and myself, and so forth, were terribly overworked. It was sometimes a question of whether we could physically stick it out; but we do feel that in these times, although we know that economy is necessary, it is not right to kill us. We were at it till 12 o'clock at night over and over again, Sunday after Sunday, Saturday after Saturday. . .you could say that the Diseases of Animals Branch staff should not be sweated. . .you cannot get first-class work out of men who are dead tired.*[26]

Conditions were no different in 1923. MAF veterinary inspector, Mr Berry, was the epidemic's first victim. His unceasing and largely unsuccessful efforts to tackle FMD in South Cheshire resulted in illness; but fearing the loss of his pension he refused to step down and eventually suffered a heart attack. Farmers believed him incompetent; but Stockman vigorously defended Berry, along with the rest of his staff: 'It has been absolute slavery for them; but we have all stood it and we are not going to complain, but we are not prepared to accept blame if our health breaks.' He later spoke of his own struggle against FMD, stating: 'I am 54 now and I suppose I look younger. I am not likely to break down now; I have had too many trials.'[27] Such confidence was sadly misplaced. Stockman, too, became a victim of FMD, dying suddenly just two years later.

During December 1923, the national press became increasingly impatient with MAF's efforts, and as the death toll and the compensation bill grew, political pressure mounted for a change in FMD control policy.[28] The government decided to appoint a special cabinet committee to consider the situation. At its first meeting on 11 December, the committee decided that slaughter should continue for the time being. However, members questioned Stockman's conviction that this was the only method of tackling FMD, and invited eminent medical scientist and director-general of the Army Medical Services, Lieutenant-General Sir William Leishman, to pass an opinion on the matter. Leishman replied that a lack of scientific knowledge about the nature and spread of FMD made it impossible to evaluate the existing policy. He went on to recommend more research into the disease, and his input encouraged the government to establish a FMD research committee, which began work during the summer of 1924 (see Chapter 5).[29]

Pressed by a deputation of Cheshire farmers, Stockman agreed to attend a meeting at Crewe on 20 December, as did Sir Francis Floud, MAF permanent secretary. There, the two men experienced a 'full and frank expression of views and considerable criticism of the methods of the ministry'. Farmers were angry and indignant at MAF's continuing failure to control FMD. They were in no mood to accept the pro-slaughter arguments put forward by Stockman, Floud and German, and instead demanded permission to isolate their stock. Stockman and Floud argued that isolation would lead to even greater hardships than slaughter. Farms would have to be quarantined for months under the surveillance of an 'army' of officials. There would be no government compensation, and the county would be ostracized by the nation. In any case, it was impossible to isolate selected cows because there was no fair means of deciding which to save and which to sacrifice. Indeed, Stockman privately believed that none were worth saving. He told the committee of inquiry: 'I do not like to say it, but in Cheshire I have never seen such a collection of bad cows in all my life.'[30] There was probably some truth in this claim, despite the local rhetoric. Bovine tuberculosis (TB) was prevalent in the area, and the lack of farming participation in MAF's milk recording scheme suggests that few Cheshire farmers were interested in improving milk production by the selective breeding of livestock.[31]

Although some farmers were swayed by Stockman's portrayal of the dangers, difficulties and costs of isolation, the more desperate among them rejected the official line. They demanded that MAF take note of the suffering inflicted by the slaughter policy and grant them the right to work towards their own salvation. They proposed a resolution: wholesale slaughter had not achieved its desired object, and therefore isolation and treatment must begin immediately, under conditions worked out by a committee of dairy farmers in consultation with MAF. It passed by 63 votes to 59, though there were numerous abstentions. The Cheshire Chamber of Agriculture and Cheshire and Shropshire Farmers' Union branches then conferred in selecting a standing committee to discuss future policy with Stockman. The majority of its members supported slaughter and they failed to agree upon an alternative policy. On hearing the news, one farmer told the local press: 'We are at a dead end, and must sit down and watch Cheshire's great dairy stock roasted in front of our eyes.'[32]

In other areas of Britain, where slaughter had succeeded in first containing and later eliminating FMD, farmers adopted an uncompromising attitude towards Cheshire's plight. They feared that under a policy of isolation, FMD would escape from the county more easily, and so insisted that the slaughter must continue. Abandoned by the nation, a profound sense of isolation settled over Cheshire. Christmas celebrations were cancelled as disease continued to rampage, and churchgoers in Chester cathedral prayed for deliverance from the plague that had befallen them. But the siege also gave rise to a sense of solidarity as farmers, practising veterinary surgeons, local authority officials, and landowner Sir Henry Tollemache joined in opposing the slaughter.[33] Also prominent in this campaign were several Cheshire medical officers of health, who condemned Stockman's 'policy of delay, obstruction and ineptitude' at farmers' meetings and in the press.[34]

Stockman genuinely sympathized with Cheshire farmers' complaints. He told the committee of inquiry:

> *When they attacked us, of course, I do not say they were not right in many cases from their point of view. They said: 'We are suffering for the good of humanity and we have suffered rather too much.' I was in absolute sympathy with them, and I would not say a hard word to anyone about that.*[35]

However, he was infuriated by the intervention of medical officials in what he regarded as a purely veterinary and agricultural affair. Doctors were, at that time, held in significantly greater esteem than vets. Their views influenced public opinion, and seriously undermined Stockman's personal and professional authority (see Chapter 4). He tried, without success, to persuade the committee of inquiry to criticize the offending individuals:

> *I do not think that should have been allowed. . .the trouble created by these official people, people of standing, was enormous during the fight. We had meeting after meeting to stop the riot, and as Sir William [Bromley Davenport, lord-lieutenant of Cheshire and member of the committee of inquiry] knows, it was a question of letting them hit at us; we simply did not bother with anything, waiting until we had this outbreak of disease settled. Now that is done, we feel we ought to tackle the men whom we think impeded our efforts. . . I think it was*

*perfectly disgraceful that they should have been allowed to do
that. I make that protest because at the time I felt it a heavy
obstacle to success, although I set my teeth and stuck it out.*[36]

A further meeting between Stockman and the farmers' committee was
scheduled at Crewe on 1 January 1924. By then, Cheshire had lost
approximately 20,000 cattle, of which 90 per cent were in calf, and in the
previous week had experienced a record 211 new outbreaks of disease. Word
spread that this would be a public meeting, and as the committee began
its deliberations, between 200 and 300 angry farmers, many of whom had
lost their stock, stormed the room in which the meeting was taking place.
German tried hastily to calm the anarchic crowd and the terrified committee withdrew into another room, leaving farmers to hold an impromptu
mass meeting. Stirring speeches were made, denouncing MAF's ignorance
of FMD and its incompetence in executing the slaughter policy. Farmers
cited numerous instances in which infected animals had recovered from
FMD before 'red tape came along and killed them'. They argued that the
official definition of isolation was unacceptable and that other, less arduous
methods of disease control were possible. A resolution demanding an end
to the slaughter was supported by all but six. Several hours later, the
committee announced the result of its deliberations. Six members had voted
in favour of slaughter and five against. It declined to take further responsibility for action and asked the Cheshire Chamber and Cheshire Farmers'
Union to arrange a vote upon the matter.

These events made national news. Even the pro-establishment *Times*
was moved by the display of local feeling, arguing that it was time for the
government to realize that the drastic enforcement of slaughter, however
much recommended as a general rule, may have ceased to make economic
sense in Cheshire. The *Crewe Chronicle* criticized the committee's 'lame
conclusion'; but the *Cheshire Observer* welcomed the novel and democratic
idea of allowing farmers to vote. All seemed confident of an overwhelming
vote against the slaughter policy.[37] The Cheshire County Diseases of
Animals Committee immediately passed a resolution supporting the
extension of isolation to dairy cattle,[38] and the Cheshire Chamber of
Agriculture similarly urged MAF to suspend the slaughter. However, to the
surprise of onlookers, 14 out of the 18 branches of the Cheshire Farmers'
Union voted against isolation. Complaints about the irregularity of the vote
soon followed. Many branches had apparently failed to differentiate
between completely replacing slaughter with isolation and continuing to
slaughter while permitting isolation in selected cases. Where this distinction
had been made, a majority favoured the latter option.[39] The inconclusive
outcome of the vote forced the decision back to the original committee,

which met on 8 January with Stockman and Floud. After a four-hour discussion, it voted upon extending isolation to selected dairy cattle in cases where both MAF and the owner agreed. The first vote was evenly split and in the second, advocates of isolation lost by one vote. This left Stockman with a free hand to continue the slaughter policy.[40]

Stockman's decision to involve local farmers in discussions about FMD control was unique to Cheshire, and reveals the tension that lay at the heart of the slaughter policy. In many ways, this was a dictatorial policy that overrode individual or local disease-control preferences in favour of uniform, compulsory and centralized action. At the same time, however, its success depended upon the voluntary actions of individual farmers. MAF officials were helpless until stockowners informed them of the appearance of FMD, and disease could not be contained unless farmers adopted 'strongly suggested' but not legislatively enforced modes of behaviour, such as confining themselves to their homes and placing disinfected straw mats outside farm gates. Legally, the policy had few 'teeth', and the absence of major deterrents to law-breaking forced MAF to depend upon the willing compliance of concerned individuals. In most parts of Britain, Stockman had managed to gain the necessary farming support by arguing that slaughter was the cheapest, most effective method of disease control, and that in restoring Britain's valued freedom from FMD, it would benefit the nation as a whole. In Cheshire, however, desperate farmers refused to accept these arguments. Their lack of support threatened the future of Stockman's preferred policy and forced him to the negotiating table.

Many of Stockman's critics were extremely pleased to hear of his meetings with the farmers' committee. They felt that MAF had at last acknowledged their suffering and accepted the validity of their complaints, and for a short period, they suspended their attack on the slaughter policy. This was not the only way in which the committee's deliberations bene-fitted Stockman. FMD control became a private rather than a public matter. 'Rank-and-file' activists could no longer dominate the agenda, while the representatives of agricultural organizations – which at a national level supported the slaughter policy – gained a greater say. Consequently, the committee agreed to let Stockman persist with the slaughter until an alternative had been decided upon. This decision was crucial because, on the ground, future chief veterinary officer P J Kelland was slowly catching up with FMD. He stopped the time-consuming process of salvaging meat from healthy contact animals, tried to slaughter the most infectious animals first, and organized agricultural students and farmers' sons into an effective labour force. He gained valuable assistance from John Done, a local farmer and prominent Cheshire Farmers' Union official, who threw his weight behind the slaughter policy, accommodated Kelland at his farm, and urged his colleagues to participate in FMD control. Not all appreciated his efforts; he was branded a 'ministry man' and ostracized by the local community. Nevertheless, in late December, his and Kelland's efforts began to pay off and the epidemic went into decline.[41]

As a result of these developments the most pressing issue for Stockman as he began meetings with the farmers' committee was not whether slaughter could control FMD, but whether it would be allowed the opportunity. He knew that most critics objected not to the slaughter policy itself but to the ineffective manner in which it had been executed. If he could only maintain it for a few more days until the falling incidence of FMD became clear, the battle would be won. The committee's decision to allow a farming vote upon FMD control bought an additional stay of execution. Voting at the various branches took over a week. Most farmers were inexperienced in politics and unaccustomed to taking decisions that affected others in addition to themselves. Faced with a highly controversial and politically important decision, and pressed by farmers' leaders to take a 'broad, unselfish' view of the situation, most proved reluctant to follow through the expressions of resistance that had characterized public meetings.[42] Fearing to step into the unknown and adopt a policy opposed outright by MAF and the NFU, they agreed that slaughter should continue, with the possible exception of selected animals.

RESOLUTION AND AFTERMATH

Several prominent opponents of slaughter refused to accept the committee's narrow rejection of partial isolation and persisted in their demands for a change in policy. Led by Sir Henry Tollemache, they formed a deputation and travelled to London to meet with Stockman. They told him that because the committee had contained representatives of neighbouring counties, it had not accurately represented the view of Cheshire farmers, most of whom supported the isolation of selected herds.[43] Realizing that the issue was not going to go away, Stockman reluctantly agreed to permit isolation, in carefully circumscribed areas only, of selected cases chosen by his officials in consultation with a farmers' committee. He warned that few animals would be eligible, there would be no compensation, and emphasized that his actions did not represent a departure from the slaughter policy.[44]

A few days later, Stockman expanded upon his decision in correspondence to J Sadler of the Cheshire Chamber, who forwarded it to the local press for publication. In his letter, Stockman expressed his hope that the 'mental strain' suffered by Cheshire farmers would not result in any 'misunderstanding' of his actions. He claimed to have been in favour of partial isolation from the very start, but had been unable to act because farmers had demanded compensation and insisted upon their right to select animals for isolation. In withdrawing these 'impossible' conditions they had cleared their own path to partial isolation, which was appropriate only in very limited circumstances. He hoped that farmers would now allow the slaughter to proceed unhindered, and would not hold any more meetings since they enabled the infection to spread.[45]

Tollemache and his colleagues bitterly rejected Stockman's claim that the full onus for disease control had always rested with farmers, who had become more enlightened over time. He and his colleagues still believed that slaughter had been a totally inappropriate response to FMD in Cheshire and that MAF's concession occurred much too late to make amends. But their influence waned as disease incidence fell during January 1924.[46] As Stockman had predicted, criticisms of the slaughter lessened. They diminished further in February, when the government announced the appointment of a scientific research committee to investigate ways of 'making FMD less harmful to Britain'. In the end, a handful of farmers benefited from MAF's policy concession. In February, Stockman offered to help farmers restock the county on condition that partial isolation cease. Local farmers' leaders, who believed that the end of the epidemic was now in sight, agreed to this proposal, and the region was returned to a fully fledged slaughter policy.[47]

There were only six cases of FMD in March, none in April and one in May. With the passing of the crisis, farmers began to look to the future. They restocked their empty farms, markets resumed and, slowly, life returned to normal.[48] In the meantime, the post-mortem began. As in 1922, the committee of inquiry into the epidemic was chaired by land-owner Captain Ernest Pretyman. Other members included Harry German of the NFU and Sir William Bromley Davenport, the lord lieutenant of Cheshire who had chaired the committee of Cheshire farmers.[49] Several representatives from Cheshire attended its hearings and submitted extremely damning evidence. Sir William Hodgson of the County Diseases of Animals Committee claimed that MAF's actions had amounted to 'an administration of despotism and tyranny which is not in consonance with the ethics of the British people'. Other witnesses argued that slaughter had succeeded only because there were no livestock left in Cheshire. They accused officials of failing to control disease promptly, of carrying the infection between farms and wrongly refusing offers of help from local authorities and veterinary surgeons. In response to this onslaught, Stock-man claimed that farmers and dealers had wilfully concealed FMD, evaded movement regulations and neglected to disinfect themselves. Also, in forcing him to defend the slaughter, they had diverted him from his true line of work, the elimination of disease.[50]

Records of the committee's proceedings show that, from the very outset, it was utterly convinced of the merits of slaughter. Pretyman told Stockman:

> *The slaughter policy is undoubtedly the only one policy. There is no difference of opinion. There is no single human being that I know of that has any common sense or any knowledge of the subject that is not in favour of the slaughter policy. . .but a half-hearted policy is a fatal thing and it is because the policy has been carried out in a rather halting manner, not by you but by the country, that it has not succeeded.*[51]

Believing that the problem lay not in the slaughter policy itself but in the manner of its execution, committee members took little heed of the pro-isolationist views of Cheshire representatives and reported that 'A policy of isolation would be equivalent to the abandonment of any hope of eradicating the disease from this country.' They clearly saw FMD control as a national problem that could only be combated effectively by a policy that placed the interests of the nation above those of the region or the individual. FMD control was the responsibility of central government, and as the state bore the cost, stockowners and local authorities had no right to select the policy.

While admitting that MAF had been disorganized and inefficient, the committee's report highlighted many other reasons for the delayed control of FMD, such as the 'indifference of the owner to the importance of prompt reporting' and the failure of local authorities to take effective precautions against disease spread. It proposed to rectify these problems by enhancing the compulsory, centralized and paternalistic nature of FMD control and minimizing the role of discretionary, individual and local decision-making.[52] Stockman welcomed these suggestions, claiming 'there is only one sound way of dealing with disease, and that is by autocratic powers. . . It must be that the powers are invested in the chief administrator, who must act at once.'[53]

On its publication in March 1925, the committee's report marked the close of a tumultuous period in the history of FMD. In echoing its predecessor's endorsement of the slaughter, this second Pretyman committee set the seal upon the three-year transformation of the slaughter policy from new and controversial measure to proven, effective method of FMD control. The events of 1922–1924 were therefore crucial in generating acceptance of a policy that was to survive into the 21st century. For a brief period, the opponents of slaughter stood a very real chance of overturning what they quite rightly viewed as an untested and inefficient policy. Had they succeeded, the history of FMD in Britain and the world would look very different. As it was, however, serious resistance to slaughter disappeared as the disease came under control and the policy finally proved its worth. Widespread criticisms did not re-emerge until 1951–1952, when in the face of widespread FMD, members of the public again called for an end to the slaughter, this time demanding its replacement with vaccination.

Although the 1923–1924 Cheshire epidemic was soon forgotten, it had tremendous historical significance. The eventual elimination of FMD and the Pretyman committee's support for the slaughter convinced officials that this measure could and should control FMD, and that they had been right to resist public pressure for a change in policy. Imbued with new confidence, they chose, during subsequent epidemics, to slaughter all infected animals, even pedigree stock, and saw little reason to negotiate with their critics. Never again did the CVO allow ordinary farmers to vote upon FMD control, and the handful of cattle preserved from slaughter in 1923–1924 were, with the exception of a few experimental animals, the last British livestock ever to recover from the disease. The fate of FMD-infected animals and their owners was sealed so that when, over 40 years later, another FMD epidemic hit Cheshire, farmers suffered as their ancestors had suffered the loss of hundreds of thousands of stock.

It is important to recognize the timeless quality of the social, financial and psychological hardships inflicted by the official FMD control policy. This chapter relates the suffering of a small rural community over 80 years ago, when both agriculture and the role of government were very different than today. But as Chapters 7 and 8 will show, witness reports dating from 1967 or 2001 strongly resembled the expressions of fear, anger, panic and sorrow related here. This is the human tragedy of FMD. It will recur as the disease recurs, for as long as the British government persists in a policy of slaughter.

Chapter 4

Effects on the Anglo–Argentine Meat Trade, 1924–1939

THE RISE OF THE INTERNATIONAL MEAT TRADE

At the turn of the 20th century, Argentina was a prosperous and peaceful nation, half way through a 50-year period of remarkable economic growth. Its most important products were livestock and agricultural goods, and its most important market was industrial Britain, which needed meat to feed the burgeoning urban population. In return, Argentina absorbed large quantities of British manufactured goods. Meanwhile, British investors built railways across the Argentine pampas, bought up meat-packing plants and shipping lines, and participated in Argentine commerce, banking, government bonds and public services. Indeed, links between the two nations were so close that Argentina was sometimes regarded as an unofficial member of the British Empire. The importance of cattle ranching to the Argentine national economy was reflected in the political dominance of an aristocratic landowning elite, which had consolidated its hold over vast tracts of land by driving out indigenous peoples. Many joined the Argentine Rural Society, and invested heavily in British pedigree animals in an attempt to improve the quality of indigenous stock.

Argentina's economic success resulted mainly from the exponential growth of its meat export trade, which was facilitated by the late 19th-century discovery of new methods of freezing and refrigerating meat. The early 1880s saw the establishment of several *frigorificos* (which were a combination of slaughterhouse, meat-packing plant and cold store) along the estuary of the River Plate. The owners, mostly British, bought livestock direct from the farmer, slaughtered them and packed the carcasses onto refrigerated ships for the voyage to Britain. Between 1890 and 1899, they sent an average 38,500 tonnes of frozen mutton and lamb, and 2800 tonnes of frozen beef to the UK each year. Meanwhile, advances in steamship design cut journey times between Argentina and England from three months to three weeks, allowing companies to ship live animals across the Atlantic for slaughter on arrival in Britain. By the later 1890s, an average 70,756 fat stock made the journey every year, and trade profits soared to record levels.

However, there was a cloud on the horizon. Foot and mouth disease (FMD) had arrived in Argentina during the late 1860s, and by 1900 it had become endemic. On learning that the disease had spread to Buenos Aires Province, where most meat and livestock export businesses were situated, the British Board of Agriculture announced that it would no longer accept Argentine livestock imports. Hoping to persuade the board to change its policy, the Argentine government passed FMD control legislation. This required the isolation of infected animals and prevented the movement of livestock off infected premises. Convinced, the board reopened the ports in 1903; but during the months that followed, its veterinary inspectors discovered several cargoes of FMD-infected Argentine cattle, and the ban was swiftly re-imposed. From then on, the board ignored repeated Argentine assurances that FMD was under control and refused to resume trade, claiming that the dead meat trade was a considerably more humane way of shipping meat to Britain. The Argentine government believed – and not without grounds – that the board still suspected its livestock of harbouring FMD. Its national pride affronted, it condemned the trade ban as politically motivated and announced in 1910 that it would henceforth only accept British livestock imports when Britain had been free of FMD for six months. British agriculturalists saw this as an unjustified retaliatory policy and lobbied vigorously for its relaxation; but it was five years before the Argentine government gave way.[1]

The Argentine government's reaction to the board's trade embargo was out of all proportion to its economic impact, as turn-of-the-century improvements in refrigeration techniques meant that livestock could be easily exported in the shape of chilled meat. This product was more perishable and required more careful handling than frozen meat, but was popular with the consumer and significantly cheaper than the fresh-killed British product. As the chilled meat trade expanded, new firms were established, some of them US owned, although the huge investment needed to found and run a *frigorífico* restricted the number of competitors. The importance of the meat trade to the Argentine national economy meant that the owners of these plants wielded considerable power. They intermittently fought for control of the market; but most of the time they maximized their profits by operating as a 'pool' that shared out shipping space and regulated exports to Britain according to demand. They also exerted some control over supply by purchasing large quantities of cattle direct from the larger ranches (*estancias*) and filling any excess shipping space with livestock bought at local markets.

During the 1910s, the Argentine meat trade continued to grow. But in 1921, by which time meat exports stood at an annual 405,000 tonnes, the post-war depression led to a sudden fall in demand, much to the alarm of cattle producers. After two years in the doldrums, trade recovered; but from then on demand fluctuated and cattle ranching became a much less secure venture than in past years. Many Argentines, nevertheless, continued to pin their hopes upon the further expansion of the meat trade, which they believed would enhance national economic development and allow their country to compete with the US for commercial leadership in South America. The most important threat to these ambitions was FMD, which continued to spread unchecked. British (and American) authorities were increasingly concerned by this state of affairs, and when, in 1926, British scientists discovered that the meat of infected animals could convey the FMD virus, the future of the Argentine meat export trade hung in the balance.[2]

The problem of preventing FMD from entering Britain in Argentine meat imports was to vex farmers, veterinary surgeons, officials and meat traders in both countries for over 40 years. The issue finally came to a head in 1967–1968 (see Chapter 7). This chapter examines early debates in the years leading up to World War II, and explores the 'spin', subterfuge and scientific manipulation that characterized the British government's response to the disease threat. As earlier chapters showed, FMD was not simply a scientific problem or a biological disease. It was also an issue of tremendous political, economic and cultural importance, and was to exert a powerful influence upon 20th-century Anglo–Argentine relations.

SUSPICIONS AROUSED?

It was during the mid 1920s that veterinary officials of the Ministry of Agriculture and Fisheries (MAF) first began to consider whether FMD could spread via infected meat. They had known for some time that virus was present in the blood during the early stages of FMD infection, and that inoculation of infected blood into susceptible animals caused disease.[3] However, new evidence obtained by the 1924 Pretyman committee of inquiry suggested that, on several occasions, pigs had contracted FMD after eating swill containing scraps of foreign meat.[4] This was a potentially serious matter: fresh-killed meat from Europe, and chilled and frozen imports from

South America, New Zealand and Australia, made up over 50 per cent of British meat supplies, and in the former two regions, FMD was rife. In a new leaflet, 'Advice to Farmers on FMD', MAF recommended that they boil swill before feeding to kill any virus.[5] Meanwhile, the chief veterinary officer (CVO), Stewart Stockman, tried to obtain more definite information on the duration of virus survival at different temperatures and in different parts of the body. This project was one of several taken up in 1924 by the newly formed FMD Research Committee (FMDRC), of which Stockman was a member. Scientists also tried, without success, to detect FMD virus in pigs' feet imported from Holland. Stockman wanted to obtain additional experimental material from diseased animals killed at the Argentine *frigorificos*. He travelled to Argentina to investigate the matter further, but died suddenly on his return.[6]

Evidence obtained during the summer of 1926 swept away all remaining doubts on the matter. MAF veterinary officials investigating a disease outbreak in cattle at Carluke, Lanarkshire, revealed that the source of infection was a nearby bacon factory, from which virus had spread either in factory effluent, which had contaminated nearby grazing pastures, or by carriage on workmen's clothes. Significantly, the factory had recently begun to import pig carcasses from the Continent for curing, and on inspection, several showed typical FMD lesions on their feet. Scientists then demonstrated that tissue samples taken from imported Dutch carcasses caused symptoms of FMD when inoculated into a susceptible cow. Here was the proof that MAF had been looking for. Shortly afterwards, Minister of Agriculture Walter Guinness informed the nation of the discovery of 'definite and complete evidence that FMD is brought into the country by foreign meat'. Claiming that further steps were needed to secure Britain's boundaries, he announced a forthcoming ban upon imports of carcasses and offal from FMD-infected European nations.[7]

At first, farming organizations were extremely pleased by the trade ban. Lord Mildmay of the Royal Agricultural Society (RAS) trumpeted: 'At last there had come the opportunity of dealing with a most dangerous channel of infection'. Criticisms came mostly from participants in the Continental trade, whose businesses faced ruin. They argued that the trade embargo was unjust, that it had halved the London meat supply and created price rises. MAF spokesmen dismissed such complaints, pointing out that the volume of Continental imports was extremely small, and that according to data collected by the National Farmers' Union (NFU), the ban had not affected the retail price of meat.[8]

Before long, however, MAF's meat import policy became the subject of a wide-ranging controversy that was played out in Parliament and in the meetings of agricultural and trade organizations. Following the imposition of the Continental trade ban, farmers and other interested parties naturally wanted to know what action MAF proposed to take against meat imports from FMD-infected South American countries.[9] Guinness replied that no action was necessary, as scientific experiments had showed only that virus could survive in fresh carcasses brought from the Continent, not in frozen carcasses from South America. In that case, asked consumers' representatives, could he allow the importation of frozen meat from the Continent? No, replied Guinness, for there was no reason to suppose that virus did *not* survive in frozen meat. But why, asked his critics, did he persist in allowing chilled meat imports from Argentina when it was known that the blood they contained could carry infection? Because, he argued, South American imports presented a lesser 'degree of danger' than European meat. FMD was less prevalent in that part of the world, control measures were more effective, and there was a greater possibility that virus would die out during the long journey to Britain. Also, there was no direct evidence that South American meat imports had ever caused a British FMD outbreak.[10]

In fact, Guinness's claims were highly questionable. Almost all South American meat imports came from Argentina, where no official statistics of FMD incidence existed and disease was reputedly widespread. MAF had no knowledge of the control measures used there, and on seeking out such information, learned that the implementation of measures introduced in 1900 and 1902 was spasmodic and evasion widespread. It was impossible to say whether Argentine meat was dangerous because scientific investigations were still in their early stages, and the feet of Argentine carcasses were commonly removed before export. However, CVO Ralph Jackson believed that sooner or later the FMD virus would be detected in Argentine meat and 'then a serious situation will have to be faced'. Also, Guinness had overstated the significance of scientists' findings. They had carried out only one inoculation experiment, and the material used had not come from carcasses contained within the Carluke factory at the time of the outbreak. Hence, there was no direct scientific proof that Continental meat imports had 'caused' a British FMD outbreak.[11] This mass of evidence shows clearly that Guinness was engaging in a cover-up operation. But what was his motivation? After all, FMD was a widely feared disease that had recently caused two of the most expensive, devastating British epidemics on record.

In short, there was far more at stake than British agriculture. As far as the British government was concerned, the interests of the consumer, the manufacturer and the investor took precedence over those of the farmer, and MAF had to adjust its agenda accordingly. Britain simply could not feed itself without Argentine meat imports. Even if domestic production and imports from FMD-free countries such as New Zealand and Australia increased, they could not match the huge volumes supplied by the Argentine trade. Interference with the latter would therefore increase the price of meat. Political agitation would surely follow, for the British valued meat consumption extremely highly, not only on nutritional grounds but also because it symbolized affluence and social status. The government had learned bitter lessons only four years before in a row over the Canadian livestock import trade. Although Canada was then free of the major contagious diseases of livestock, the agricultural lobby and minister of agriculture, Sir Arthur Griffith-Boscawen, opposed the relaxation of a trade embargo imposed 30 years earlier when bovine pleuro-pneumonia was discovered in Canadian cattle. Critics such as the *Daily Express*, which ran a massive campaign in support of trade resumption, accused the government of depriving the population of cheap meat. Newly enfranchised urban voters expressed their displeasure by turning Boscawen out of his Dudley seat at the next by-election. The message was clear: don't meddle with the people's food.

There was another important reason why the British government refused to ban Argentine meat imports. It knew that the meat trade was of tremendous economic importance to Argentina and was heavily invested with national pride. For these reasons, the Argentine government was likely to retaliate – as it had in 1910 – against British imposed trade restrictions. Such measures would also harm British investors, who had substantial stakes in the *frigorificos*, railways and shipping. On balance, therefore, the meat trade had to continue, despite the threat that it posed to British agriculture.[12]

The fact that agriculture was awarded so low a priority by the British government was indicative of its declining influence and perceived irrelevance to national prosperity. The drift away from the land, which had begun during the early years of the Industrial Revolution, continued during the late 19th and early 20th century. Agriculture – especially the arable sector – fell into deep depression as foreign countries that were better equipped to produce cheap food exported surpluses to Britain and, in exchange, absorbed large quantities of British manufacturing goods. By 1913, agriculture contributed only 6.4 per cent of gross domestic product (GDP) compared to 18.4 per cent in 1856. Many landlords were forced to reduce rents in order to retain tenant farmers, or to sell off parcels of land. As their income dropped, they spent less on their estates; land was poorly maintained, and buildings and machinery fell into disrepair. Lloyd George's 1909 budget, which introduced new taxes targeted at the landowning classes, further eroded their wealth. There was a brief revival of agricultural fortunes during World War I, when the government realized, belatedly, the crucial importance of increasing domestic food production to make up for lost imports. But in 1919, in a policy U-turn later termed the 'great betrayal', it repealed wartime legislation that had guaranteed farmers fixed prices for corn. Food imports resumed and British agriculture once again sank into depression. During the 1920s, a sizeable number of agrarian Tory members of parliament (MPs) lobbied Parliament on behalf of the rural interest, while the NFU began to assert itself politically. But divided priorities and political infighting meant that they failed to capture the Parliamentary agenda; and at a time when Britain's economy was failing, exports were contracting and unemployment was rising, agriculture remained low down on the government's list of priorities.[13]

Most of the social, political and commercial factors that shaped government attitudes towards the Argentine meat trade were not explained to the public. MAF spokesmen did occasionally admit to the impossibility of depriving the public of so important a source of meat, but the majority of their pronouncements perpetuated the myth that policy was guided by science. However, officials were well aware of the contradictions inherent in the government's meat import policy. Behind the scenes, they tried hurriedly to establish some sort of disease control framework that would support the minister's claims about the safety of Argentine meat. The CVO, Ralph Jackson, met with Juan Richelet, the veterinary attaché to the Argentine embassy in London, and Colonel Dunlop Young, a veterinary expert on the meat trade who had recently visited Argentina. After compiling background information upon the nature of the Argentine meat trade and the scale of the FMD problem, the men decided that the best method of reducing the FMD risk to Britain was for the Argentine government to introduce new trade regulations. These should prevent diseased animals from leaving the *estancias*, and prohibit their slaughter for export. MAF should also station one or two veterinary inspectors permanently in the region to check that these measures worked properly.

Initially, the Argentine response to these suggestions was not encouraging. The ambassador to London, Don Evanisto Uriburu, argued that there was no evidence that Argentine chilled meat could convey infection, that effective FMD controls were already in place and that, in any case, FMD was a mild, unproblematic disease. Jackson told him that FMD was widely feared in Britain and that farmers were pressing for an outright ban on meat imports from infected countries. If Argentina did not take action and new evidence emerged to show that virus could survive in chilled meat, the British government could no longer make the case for treating the Argentine and Continental meat trades differently. This argument had the desired effect. In September 1926, the Argentine government agreed in principle to MAF's suggestions on condition that they were also accepted by other meat-exporting South American nations. MAF was delighted, describing the agreement as 'a unique opportunity by which Great Britain protects itself at the expense of the exporting country'. Then, acting through the Foreign Office, it opened negotiations with the governments of Uruguay, Brazil and Chile.[14]

Before these arrangements could be formalized, British scientists announced new discoveries that heightened the controversy surrounding the Argentine meat trade. Their experiments, performed upon the carcasses of artificially infected livestock, revealed that in beef and bacon, prepared and stored under standard commercial conditions, the FMD virus survived up to 87 days in the bone marrow and 46 days in the blood. When bones from infected animals were crushed and fed to pigs, they caused abrasions to the mouth and gums, through which virus present in the bone marrow could enter the body and cause disease. These results, which were published in February 1927 and June 1928, showed unequivocally that if infected animals were slaughtered in Argentina and exported to Britain, virus could easily survive the journey and initiate outbreaks of FMD via the feeding of pigswill.[15] They therefore undermined Guinness's claim that Argentine meat imports were far less dangerous than those from the Continent

In response to this news, MAF passed the FMD (Boiling of Animal Foodstuffs) Order, which required swill to be boiled for one hour. This, it claimed, would ensure that in the unlikely event of infected carcasses being imported, they could not initiate infection. Officials also argued that the new findings proved that they had been right to ban the Continental trade.[16] The director of the Dutch State Veterinary Service disagreed. He was still aggrieved by MAF's trade embargo, and to prove its lack of justification he cited German and Dutch scientific findings. These showed that pigs fed with the flesh of FMD infected animals did not contract FMD, and that FMD virus was killed rapidly by the rise in muscle acidity that occurred after death.[17] The FMDRC thought these results insignificant because even if virus did not survive in meat, it could endure for many weeks in the bone marrow, offal and lymph nodes. In any case, argued Guinness, the definite proven link between British FMD outbreaks and imported Dutch carcasses meant that no amount of laboratory-derived evidence could persuade him to reopen the ports.[18]

MAF's critics again drew attention to the inconsistency of its actions. Why did it require a trade embargo to stop Continental imports of virus, but only swill-boiling to prevent Argentine meat from causing disease? Surely, argued farming spokesmen, the government should take heed of recent scientific findings and extend its ban to cover Argentine meat imports.[19] On the contrary, claimed representatives of the meat trade, the government should permit the Continental trade to resume providing countries adopted the same administrative measures as Argentina. Again, Guinness tried to deflect criticisms by highlighting the proven link between

Continental meat imports and British FMD.[20] But leading agriculturalists refused to accept his explanation, arguing that the government's policy was confused, contradictory and downright wrong. Even if swill-boiling was successful – which was unlikely, as it was impossible to enforce – there were many other potential routes of disease spread. Humans who came into contact with infected meat could carry the virus upon their clothes and boots, dogs could carry infected bones into farmyards, and pigs could feed off rubbish heaps containing scraps of infected meat and bones. They also doubted whether the Argentine government could be trusted to carry out the proposed administrative measures. In any case, it was not enough to reduce the disease risk, because even one case of FMD importation could have a devastating effect on British agriculture. Instead, risk must be removed altogether by means of a trade ban.[21]

Some historians believe that British farmers only supported the banning of Argentine meat imports because they knew it would result in higher prices for home-produced meat.[22] In certain cases, this was undoubtedly true. For example, the Council of Agriculture for England supported a resolution put forward by George Courthope which stated that, due to the survival of FMD virus in the bone marrow, all chilled Argentine meat imports should be boned. Courthope later admitted that his aim was to force the Argentines to freeze all their meat exports, thereby eliminating trade competition between home-killed and Argentine chilled meat.[23] Supported by the RAS, former President of the Board of Agriculture Lord Ernle suggested that MAF place all chilled Argentine meat imports in cold storage for three weeks after arrival in order to kill the virus. This would have the additional effect of ensuring that Argentine meat looked far less palatable than the British variety.[24] However, British farmers' demands were not entirely self-interested and arose partly from their very real fear of FMD. Many had suffered appalling hardship during the epidemics of 1922–1924 and when, after several years' lull, FMD incidence increased during the early months of 1928, they were extremely concerned that another epidemic was just around the corner. Like the vast majority of the population, they had great faith in scientific enquiry, and believed that recent discoveries demonstrated a real risk of FMD invasion. And in highlighting the multiple and indirect routes of FMD spread, they were only echoing arguments used by MAF officials to justify their earlier embargo upon Irish livestock imports.

Guinness and his colleagues tried to deflect criticism by claiming that scientists had merely demonstrated the 'theoretical' possibility that Argentine meat could convey FMD virus, and that this should be distinguished from the actual experience of viral carriage by Continental carcasses. But, at the same time, he argued that it was not necessary to search for a more definite link between Argentine meat imports and British FMD cases – for example, by feeding Argentine meat to pigs and waiting to see if they developed symptoms – as this would merely 'test the existence of a contingency which has already been admitted and which has already been provided against'.[25] Even if germs were imported in Argentine meat, they were not dangerous unless they gained access to susceptible livestock. The illegal feeding of unboiled swill was the only way in which this could happen; all other postulated routes of virus transmission were mere 'speculation'.[26] Such claims not only contradicted the commonly accepted view of FMD as a highly contagious disease that was capable of spreading via multiple and indirect routes. They also portrayed scientific research as an activity that had little or no relation to the real world, and which should certainly not form the basis for policy decisions.

THE ARGENTINE REACTION

Meanwhile, trouble was brewing in Argentina over the introduction of the new FMD control measures. In September 1927, J L Frood, MAF veterinary adviser in Buenos Aires, reported that the Argentine authorities had insufficient resources to implement controls, and that opposition was widespread and evasion commonplace. Some parties, such as the British newspaper *Review of the River Plate* thought that the new regulations were necessary to counteract British farmers' calls for a complete trade ban. However, many cattle producers, *frigorifico* owners and politicians argued that they were wholly unjustified and resisted their imposition. They claimed that FMD was only a mild, insignificant ailment, that existing FMD controls were completely effective and that all Argentine livestock were healthy. Having assisted Jackson in framing the new regulations, Juan Richelet, Argentine veterinary attaché, now claimed that his country never exported diseased meat because of its system of veterinary meat inspection, which was among the best in the world.

Many Argentine critics believed that the British government's drive for additional FMD controls was a political act. They knew that moves were afoot in Britain to replace the free trade policy that had prevailed since Victorian times with a new policy of imperial preference, which would levy duties on imports from foreign countries and grant preference to goods from the empire and dominions. On learning that meat exporters such as New Zealand and Australia were not required to adopt the same trade regulations as South American countries, they accused the British government of using FMD as a smokescreen to disguise its introduction of a new trade policy.[27]

At home, Guinness and his MAF colleagues played down the significance of British scientists' findings in order to deflect calls for a meat import
ban. But they played up their importance during negotiations with
Argentine officials who were, as yet, unconvinced of the need for new trade
regulations. Unfortunately, this tactic backfired because the US was using
the same scientific evidence to justify an outright ban on Argentine meat
imports.[28] Like Britain, the US pursued a goal of national freedom from
FMD. It applied a similar set of preventative and control measures, and
had suffered only occasional epidemics, in 1870, 1880, 1884, 1902, 1908
and 1914. In 1924, a devastating eight-month-long FMD epidemic struck
California, resulting in the slaughter of over 100,000 animals at a cost of
approximately US$7 million. In its aftermath, the government introduced
a package of additional control measures, including a ban on meat imports
from FMD-infected regions, which came into operation in January 1927.

In contrast to Britain, the US produced enough meat to feed its
people, and so it imported only small volumes from Argentina. Nevertheless, Argentine politicians, cattle producers and *frigorífico* owners vehemently resisted its trade embargo on account of the 'diseased' stigma that
it attached to their animals. MAF's refusal to follow suit provoked outrage
among British farmers, and led Argentine interests to assume that the
scientific evidence of virus survival in meat was unsubstantiated. Guinness's
parliamentary speeches lent support to this view by referring disparagingly
to scientific 'theory'. As seen earlier, Dutch and German scientists also
disputed the significance of the FMDRC's findings, as did Argentine
veterinary scientist, Dr José Lignieres. Referring to the recent discovery of
three types of FMD virus, A, O and C, Lignieres declared that the virus
present in Britain was an O type. It could not, therefore, have originated
in Argentina, where, according to his test results, the A type prevailed (the
FMDRC later retyped his samples and deemed this conclusion invalid).
Meanwhile, prominent French FMD expert Professor Vallée told the
Argentine Rural Society that as meat was only one of many ways by which
FMD could enter and spread throughout a nation, its significance should
not be overstated.[29]

Some authors have argued that the Argentine failure to recognize
FMD as a terrible plague was due to ignorance.[30] This is unfair. While the
Argentine understanding of FMD was very different from that which
prevailed in Britain, it was not irrational. As we have already seen, the main
reason why British farmers dreaded FMD was because of the hardships
inflicted by measures used to control the disease. In Argentina, there were
no such controls and therefore the disease had very different social and
economic implications. It also presented a far milder clinical picture. FMD
symptoms were not particularly severe in fattening stock when compared

to breeding or dairy cattle, and frequent bouts of infection would have left Argentine cattle with a high degree of immunity. Furthermore, most Argentine animals were reared extensively on vast *estancias*, where they were handled infrequently and their condition monitored only occasionally. Consequently, the effects of FMD were not closely observed. So although caused by the same virus, FMD was actually a very different disease in Britain and Argentina. Indeed, after reading British descriptions of FMD, many Argentines simply did not believe that it was the same as the mild ailment that they called *aftosa*. Most thought that trying to control FMD would cause greater financial losses than the disease itself. And, living in a country where FMD was endemic and where infection spread mostly via the movements of infected stock, they simply did not understand the British fixation with disease importation. Nor did Vallée, who came from France, where FMD was similarly endemic. It is unsurprising, therefore, that the Argentines adopted him as their scientific 'champion' in opposition to the FMDRC.

Many British critics claimed that the Argentine government only denied that FMD was a problem and refused to commit to disease eradication because it wanted to maintain the meat export trade and to avoid imposing unpopular domestic trade restrictions. To a certain extent this was true. An export ban would have seriously undermined national morale, damaged the economy and affected the government's political standing both at home and abroad. Domestic trade regulations were also problematic. Influential cattle producers were angered by the way in which they treated each vast *estancia* as a single unit, with no livestock movements being permitted for 21 days after FMD appeared. Powerful *frigorifico* owners objected to the expense, since disease controls interfered with the supply of livestock and required infected carcasses – of whatever quality – to be sold cheaply for local consumption or as canned meat.[31] If, indeed, FMD was a far milder disease in Argentina than in Britain, one can hardly blame Argentine politicians, livestock owners and businessmen for believing that control was more costly than the disease itself. And it must not be forgotten that British attitudes towards FMD were similarly shaped by political and economic motives.

Pressurized by the British government and well aware of British farmers' demands for a complete trade ban, the Argentine government overrode popular resistance and passed a decree, in February 1928, that laid down new restrictions upon the movement and slaughter of FMD-infected animals. MAF feared that the powerful opposition of cattle producers and *frigorifico* owners would mean that, like so much other South American legislation, the decree would not be enforced. It decided, therefore, to station a veterinary inspector in Argentina to check that measures were properly implemented, and suggested that Parliamentary Secretary Lord Bledisloe (formerly Charles Bathurst) take time out of a planned pleasure cruise to explain the new regulations to Argentine cattle producers and encourage their enforcement by the state. During his visit, Bledisloe learned that there were few systematic attempts to prevent the entry of FMD-infected animals into the *frigorificos* or to halt their export to Britain. He likened the situation to that of Britain 60 years previously, and argued that imposing strict controls in a nation where FMD was generally disregarded would only lead to evasion. Instead, he asked *estancieros* to make it 'a matter of honour and conscience' not to move diseased stock and to teach other cattle producers about the British fear of FMD. This, he believed, would make an impression upon the 'non-European mind.' His visit quelled Argentine fears of an imminent trade ban, won over the influential Argentine Rural Society and led to the signing of the 'Bledisloe agreement', a document that laid down the conditions under which Argentina agreed to export livestock to Britain.[32]

The Bledisloe agreement was a huge relief to MAF. Finally, and with confidence, it could state that measures were in place to stop the export of diseased livestock from Argentina to Britain. It also warned farmers that if they did not stop agitating for consistent treatment of Continental and South American meat trades, it would be forced to lift the Continental trade embargo. This rhetoric, together with the falling incidence of FMD during the late 1920s and early 1930s, dampened British farmers' demands for a trade ban. British farmers' anxieties also diminished following the visit of Harry German, president of the NFU Meat and Livestock Committee, to Argentina in October 1928. He reported that while infected animals did enter the *frigorificos*, they were usually detected prior to slaughter, and that post-mortem inspection provided an additional safeguard against the shipping of diseased carcasses to Britain.[33]

TWISTED SCIENCE

Meanwhile, in the laboratory, investigations into virus survival in animal carcasses continued. In 1930, MAF's chief scientific adviser, Daniel Hall, requested the FMDRC to begin feeding experiments of a type which Guinness had previously rejected as worthless. Presumably because of the earlier furore over its meat import policy, tests were carried out in absolute secrecy, although the Argentine government was informed.[34] Once a week, Argentine cattle bones were removed from Smithfield market, crushed and fed to three pigs. Samples were also inoculated into guinea pigs, a species whose susceptibility to FMD infection had been recently demonstrated by

German researcher Otto Waldmann. Over the next 14 days, animals were observed for signs of disease. Then, in order to rule out the possibility that the bones had contained an extremely mild form of virus which had not produced noticeable disease symptoms but had, nevertheless, given rise to disease immunity, all animals were inoculated with the FMD virus. The presence of symptoms showed that animals had not previously encountered the disease, while their absence suggested prior infection by virus contained within the bones. Throughout these tests, precautions were taken to prevent extraneous FMD virus from accidentally contaminating bones or infecting experimental animals.[35]

Initially, all results proved negative; but in August 1930, two out of three pigs failed to develop FMD symptoms after virus inoculation. They had not shown signs of disease during the 14-day observation period, although guinea pigs inoculated with the same bone extract had done so. Paradoxically, when scientists inoculated samples of tissue and blister fluid from the infected guinea pigs into a batch of immune guinea pigs, they also developed disease symptoms.[36] However, they paid little heed to this part of the test, knowing from experience that inoculation experiments sometimes gave rise to ambiguous results. They also knew that some viruses varied in their tendency to infect different species and in the severity of disease symptoms produced.[37] They therefore decided that virus had, indeed, been present in the Argentine bones from Smithfield market and had, on feeding, induced sub-clinical FMD infection in pigs.[38]

The FMDRC (of which two prominent members – the CVO and the head of the Weybridge Veterinary Laboratory, W H Andrews – were MAF employees) interpreted the results in a very different way. They decided that the experimental animals had been accidentally infected with a virus that had recently been tested within the laboratory and had also behaved oddly when inoculated into immunized guinea pigs.[39] This conclusion was certainly viable as tests were occasionally hampered by the accidental spread of infection;[40] but it represented a rather obscure reading of experimental results and smacked of political convenience. Keen to shore up the existing FMD control policy, MAF representatives exploited experimental uncertainties in order to obtain a conclusion that suited their political and economic ends.

Subsequent events provide additional evidence for the committee's political motives. Feeding experiments continued into 1931, but no more positive results were obtained. In December 1932, Jackson's successor as CVO, John Kelland, told the FMDRC of the 'official position' upon these tests. He claimed that no definite conclusion could be drawn from the positive reactions that had been obtained in guinea pigs, and that 'if any [parliamentary] questions arose in the future, the reply should merely state that swine had been fed and inoculated with material from imported bones of South American origin but that no evidence of infection had been obtained by this method'. The committee agreed.[41] The following November, Kelland argued that feeding experiments should be excluded from the committee's next published progress report. Andrews led the committee in agreement, stating that as the results were largely negative, they could prove misleading, and that it was 'desirable to avoid giving a handle to those whose interests might lead them to draw unwarrantable conclusions in their own favour'.[42]

Meanwhile, it became increasingly obvious to MAF officials that Argentina was still exporting FMD-infected meat, and that British FMD outbreaks were occurring as a result. According to J R Frood, MAF's veterinary inspector in Argentina, the Bledisloe agreement was not working properly. There were still no national figures of disease incidence; many livestock owners neglected to notify the authorities of FMD outbreaks; railway trucks were not cleaned properly after the unloading of cattle; and FMD appeared frequently in municipal markets and almost daily in the Buenos Aires *frigorificos*. During the early 1930s, Frood reported that compliance was improving and disease incidence diminishing. However, in 1932, the economic situation warranted his recall. Deprived of a reliable source of information, MAF's only means of keeping a check on the Argentine FMD situation was to ask the Argentine government, repeatedly, to confirm its commitment to the Bledisoe agreement.[43] Three years later, in response to renewed NFU pressure for import restrictions, it sent veterinary surgeon Vincent Boyle upon a year-long tour of Argentina. Boyle discovered that FMD-infected animals were sometimes slaughtered for export to Britain; but on receipt of this news, MAF did little other than request additional assurances from the Argentine government.[44]

At the same time, MAF veterinary inspectors investigating new British outbreaks of FMD found that Argentine meat was strongly implicated in a large proportion of cases. During the five-year period, 1930–1934, they attributed 64 outbreaks to fresh invasions of virus. Of these 64 'primary' outbreaks (most of which spread to cause 'secondary' outbreaks), 42 were caused by infected imported meat, while another 18 – officially classified as 'obscure' – were possibly linked to this source. The evidence involved was purely circumstantial: diseased pigs might have consumed uncooked swill containing foreign meat; gained access to foreign meat bones; scavenged off rubbish dumps containing meat scraps or wrappers; or eaten food transported in vehicles or containers that had previously contained swill or Argentine meat.[45] In 1936, the new CVO, Daniel Cabot, circulated an internal MAF memo which stated that 25 per cent of all primary outbreaks occurring during the previous ten years were definitely caused by imported meat, and another 25 per cent were possibly linked to the trade. He also noted that FMD would recur for as long as meat was imported from infected countries. The following year, Permanent Secretary Donald Vandepeer admitted to fellow MAF officials that 'some degree of risk is therefore inevitable if we continue (as we must) to import large quantities of meat from South America. The policy has been to decrease this risk to a minimum.'[46]

But while MAF officials privately acknowledged that their meat import policy was failing to keep FMD out of Britain, they continued, publicly, to deny that there was any proven link between Argentine meat and British FMD outbreaks and to state that they were satisfied with Argentine FMD controls.[47] In order to ensure that their private suspicions did not enter the public domain, they not only censored the FMDRC's experimental reports, but also marked as 'confidential' all FMDRC papers containing details of FMD outbreaks caused by imported meat. And when farmers, politicians or journalists brought forward circumstantial evidence linking Argentine meat to British cases of FMD, MAF officials dismissed their statements as 'inconclusive', and claimed that policy should be based on fact, not suspicion.

The 1930s did, in fact, see reductions in the volume of meat exported from Argentina to Britain; but this had nothing to do with FMD. Faced with a deepening economic crisis, the British government finally abandoned its century-long predilection for free trade in favour of tariffs and trade barriers, which protected British produce from competition and granted exports from the empire and dominions preference over goods from foreign countries. By rights, this new policy should have significantly curtailed Argentine meat imports.[48] However, to the anger of British farmers, who were already suffering from 'phenomenal' falls in meat prices, negotiations with the Argentine government resulted in the signing of the Roca-Runciman agreement, which obliged Britain to maintain Argentine chilled meat imports at the same level as 1932 in exchange for reduced tariffs upon British exports to Argentina.[49] Knowing that whatever happened, they were now unable to impose additional restrictions upon Argentine meat in the name of FMD control, MAF officials lost interest in the matter and instead diverted their attention towards alternative routes of FMD invasion, such as birds, wind and wild mammals.

Shortly afterwards, the outbreak of World War II brought farmers' grumblings to an end. After all, they could hardly demand reductions in food imports at a time when German U-boat bombing in the Atlantic imperilled the British food supply. For the same reason, MAF officials could get away with admitting publicly, for the first time, that South American meat imports were implicated in many British FMD outbreaks.

But as later chapters reveal, the problem of infected meat did not go away. During the post-war years, British and Argentine governments and farmers renewed their squabbles over the Argentine meat trade and upon the nature and implications of scientific 'proof'.

Chapter 5

The Science, 1912–1958

THE BIRTH OF SCIENTIFIC MEDICINE?

To many historians of medicine, the germ theory of disease is one of the greatest discoveries of modern science.[1] Postulated by Pasteur and Koch during the late 19th century, it is supposed to have brought enlightenment and progress to a medical field still steeped in ignorance and superstition. A new science was born, bacteriology, as well as a new group of laboratory-based research workers, the 'microbe-hunters', who identified the germs responsible for disease and sought new ways of controlling them. In discovering vaccines and serums with which to prevent infection they vanquished ailments such as typhoid, plague, cholera and diphtheria, which for centuries had struck fear into the hearts of men and women. And in expanding the knowledge and understanding of disease, they made medicine 'modern' and 'scientific'. It was therefore inevitable, so the story goes, that doctors and the state would take note of their findings and encourage further research, for only the ignorant, old-fashioned or self-interested could fail to recognize the benefits of bacteriology.

Unfortunately, while appealing, this heroic tale of progress is simply not true. For one thing, rational medicine was not suddenly born at the end of the 19th century. As Chapter 1 showed, highly sophisticated disease theories existed long before that date and led to the evolution of policies that markedly improved public health. Doctors may have used different remedies back then, but to them and their patients, they worked. And veterinary surgeons without any real knowledge of germs managed to stamp out cattle plague and sheep-pox by restricting livestock movements and subjecting infected animals to slaughter or quarantine. It is also mistaken to view as inevitable the adoption and application of the germ theory. In fact, different professional groups responded in different ways on account of their varying interests and resources. Many late 19th- and early 20th-century doctors, especially clinicians, felt that the utility of laboratory science had been overstated because vaccines and serums – the supposed 'miracles' of scientific medicine – were only available against a very limited range of ailments and did not always work. Besides, they were unconvinced by scientists' attempts to reduce whole patients, as seen at the bedside, to collections of cells, tissues or animal bodies manipulated within the laboratory. Divisions opened up between different sections of the medical profession as they vied for 'expert' status and for the right to define the relationship between laboratory science and medical practice.[2]

So germ theory and the associated science of bacteriology did not suddenly alter the face of medicine. Changes occurred gradually over several decades, and it was not until the inter-war period that serums and vaccines, laboratory-based disease diagnosis and bacteriological research became integral to the medical field. Important developments included the turn-of-the-century founding of new privately and publicly funded research laboratories such as the Jenner Institute (established in 1891 and later renamed the Lister Institute), the Wellcome Physiological Research Laboratories (established in 1894 by Burroughs, Wellcome and Co, the pharmaceutical company) and the London School of Hygiene and Tropical Medicine (established in 1902, with the assistance of the Colonial Office). By the 1910s, medical schools attached to British universities had begun to offer training in basic medical science and opportunities for employment in medical research. General hospitals increasingly incorporated laboratories for the diagnosis and investigation of disease and the preparation of therapies, and local authorities began to employ medical scientists to diagnose disease in their new public health laboratories. With the passage of the 1911 National Insurance Act, more funds became available for medical research. This money was administered by a new body, the Medical Research Committee, headed by physiologist Walter Morley Fletcher. It was reconstituted as the Medical Research Council (MRC) after World War I, founded a laboratory, the National Institute for Medical Research (NIMR) and, under Fletcher's direction, became extremely influential in the funding and development of medical science.[3]

British veterinary surgeons were quick to adopt the germ theory, primarily because it justified their penchant for stamping out contagious animal diseases. However, when compared to the medical field, the new science of bacteriology had relatively little impact upon veterinary research, education and practice. The first veterinary surgeon to take an extensive interest in the subject was Professor (later Sir) John McFadyean, who had also trained as a doctor. In 1892 he was appointed to a new chair in pathology and bacteriology at the Royal Veterinary College (RVC), and with the aid of the Royal Agricultural Society (RAS) he founded a small veterinary laboratory at the school, the first of its kind in Britain.[4] Investigations under the veterinary department of the Board of Agriculture began in 1905, when McFadyean's son-in-law, Stewart Stockman, became chief veterinary officer (CVO). He rented a laboratory and appointed a committee to research into a cattle disease known as contagious abortion (or brucellosis). In the years that followed, Stockman's laboratory expanded, and by 1909 he had five assistants and was undertaking diagnostic work, as well as investigations into other animal diseases. Additionally, in 1909, the government passed the Development Act, which aimed to revitalize agriculture by increasing the money available for agricultural and veterinary research. Funding was to be administered by a specially appointed body, the Development Commission. One of its first projects was to fund a new veterinary department laboratory at Weybridge, where the Veterinary Laboratories Agency (VLA) of the Department of the Environment, Food and Rural Affairs (DEFRA) is still situated today.[5]

For almost two decades, Stockman's and McFadyean's laboratories were the only sites of veterinary research in Britain. The work that went on there had relatively little impact upon veterinary practice because most veterinary surgeons and farmers viewed slaughter as an acceptable solution to many ills, and saw little point in seeking out new methods of disease diagnosis and therapy when they would probably cost more than the patient was worth. This state of affairs was self-perpetuating because of the privately funded nature of British veterinary education. Schools responded to market demand by educating students to become veterinary practitioners. Because a rigorous scientific training was deemed unnecessary for this role, qualified veterinarians had little time for laboratory enquiries and were ill-equipped to enter the research field. Nor were Ministry of Agriculture and Fisheries (MAF) veterinary surgeons concerned about the lack of veterinary research. They knew that their predecessors had managed to stamp out diseases without recourse to the laboratory, and favourably compared their achievements to those of doctors, whose serums and vaccines had yet to eliminate a single human ailment from Britain. So, for most members of the profession, bacteriological work was irrelevant to the goals and practice of veterinary medicine. This attitude profoundly influenced the initiation and evolution of foot and mouth disease (FMD) research in Britain.

Starting Out: FMD Research in Britain and Europe up to 1924

FMD research began on the Continent in 1897, when concerns about the costly impact of endemic FMD upon meat and milk production led the

Prussian government to instruct bacteriologist Friederich Loeffler to seek a method of combating the disease. Loeffler had previously worked with Koch and had identified the germ responsible for diphtheria infection. He quickly discovered that the blood (or, more specifically, the serum) of animals that had recovered from FMD could, on inoculation, protect susceptible livestock from infection. He also realized that disease was caused not by a bacterium, but by a much smaller microbe that could not be seen through a microscope (at least not until the 1930s, when electron microscopes were discovered). He called this germ a 'filter-passing virus' on account of its propensity to pass through filters that retained bacteria. Other viruses identified in the late 19th century included the tobacco mosaic virus, the myxomatosis virus and the African horse sickness virus, discovered by McFadyean in 1899. Unlike bacteria, they could not be cultured in-vitro, and this posed problems for vaccine production. Normally, scientists made vaccines by culturing bacteria on an agar plate and then modifying (attenuating) them in some way, so that on inoculation into susceptible animals or humans they stimulated antibody formation but did not cause serious disease. Because viruses could only be grown in-vivo (inside the animal body), vaccine production was more difficult, although not impossible – as Pasteur had demonstrated in 1884 when he produced a rabies vaccine from rabbits' spinal cords.[6]

German, French, Dutch and Italian scientists built upon Loeffler's early findings, describing their work in papers published in scientific journals. They discovered ways of producing FMD serum more cheaply and in larger volumes, and found that it induced variable degrees of immunity that lasted approximately ten days. They showed that if the FMD virus was administered at the same time as serum, animals developed very mild disease symptoms but gained far longer-lasting immunity to infection. They noted also that when animals already infected with FMD were treated with serum, the severity of their symptoms and the associated losses in meat and milk production declined. Excited by these findings, several European governments established serum-production institutes. By the early 1920s, serum administration – either alone or in combination with FMD virus – was reportedly widespread on the Continent. While it had little effect upon disease incidence, its true benefits lay in reducing the economic costs of infection.[7]

FMD research began much later in Britain than on the Continent. As Chapter 1 described, endemic FMD disappeared in 1886. For the next 25 years, FMD invasions were occasional and short lived; but in 1911 it reappeared and spread to cause 25 outbreaks. Alarmed, the Board of Agriculture appointed a departmental committee to consider additional preventative measures. To several members, it seemed a propitious time to begin FMD research: no investigations had taken place since the small-scale, largely unsuccessful efforts of the late 1870s and early 1880s (see Chapter 1); the government seemed increasingly keen to fund veterinary research; new investigative techniques were now available; and the recent discovery of vaccines, serums and antitoxins against human diseases aroused hopes of finding a similar method to combat FMD. The committee therefore asked witnesses how they felt about FMD research. Many were overwhelmingly opposed, and the most vocal critic of all was Stockman, who insisted that the risks involved far outweighed the likely benefits.

Stockman assured the committee that he was not opposed to FMD research *per se*; indeed, he had a great interest in the disease ('Nothing would give me greater pleasure. . . I thirst to investigate the thing').[8] However, he believed that investigations could prove extremely dangerous. The FMD virus was highly contagious, and as French and German investigators had recently discovered to their cost, it could be easily carried out of the laboratory on clothes and shoes. For this reason, he refused to permit independent British investigations into FMD. But even if the board took charge of research and managed to contain the virus within the laboratory, its very presence within Britain would prevent the nation from attaining the 'FMD-free' status needed to export livestock to disease-free nations. In any case, Stockman argued, there was no real need to search for a FMD serum or vaccine within the laboratory because the existing policy of slaughter and import restriction was more than capable of controlling the disease. And while experiments could possibly detect whether virus was contained within imported goods or carried by the wind, birds or wild mammals, this was not a priority because it would be difficult to prevent disease from spreading by these routes, and obtaining positive results would be like looking for a needle in a haystack. Convinced by these arguments, the committee recommended in 1912 that future investigations take place outside the mainland, under conditions that satisfied the Board of Agriculture.[9]

Shortly afterwards, Stockman sent several veterinary surgeons to the British colony of India to carry out investigations into FMD. There, at least, experiments could not pose a threat to British 'FMD freedom'. However, members of the team soon discovered that indigenous Indian livestock were naturally resistant to infection. Unable to obtain meaningful results, they headed home and recommended that future investigations take place on an island off Europe.[10] Stockman interpreted this suggestion in rather imaginative fashion and arranged, in 1920, for tests to take place upon a disused warship, the *HMS Dahlia*, moored off the coast of Harwich. He appointed a small research committee of doctors and veterinarians to oversee the project, and asked Joseph Arkwright, assistant bacteriologist at the Lister Institute, to devise and carry out the experiments. Under the committee's direction, Arkwright tried to culture FMD (a feat not actually accomplished until 1929) and to devise a practical method of immunization. He also explored how different environmental conditions might affect virus survival. Almost immediately the project ran into trouble. First, the ship's crew refused to work until their pay was brought into line with union rates. Then Arkwright discovered that onboard livestock accommodation was inadequate, and delays set in while the renamed Ministry of Agriculture fitted out two extra boats. Three months later, a new threat emerged as the deteriorating post-war financial situation brought about a 20 per cent cut in MAF's 1922–1923 budget. The committee wanted research to continue; but Arkwright had had enough and resigned. His lack of progress led members to recommend that future research take place on dry land.[11]

There are several explanations for the marked difference between British and European attitudes to FMD research. Most significantly, as an island, Britain could impose stringent regulations upon its import trade. It therefore suffered only occasional invasions of FMD that could be stamped out quickly using slaughter. These measures worked well during the 1890s and 1900s, and so Stockman's veterinary department saw little point in hunting for alternatives within the laboratory, especially given the cost of research – in terms of lost exports to FMD-free nations – and the associated risk of virus escape. In Europe, however, most nations were unable to adopt British-style FMD controls. Their geography was very different, and although import regulations existed on paper, it was practically impossible to police land borders effectively. FMD therefore invaded repeatedly and became endemic.[12] The high incidence of infection made slaughter unfeasible, and livestock owners – who, like their British counterparts in the 19th century, were accustomed to, and even accepting of, FMD – objected to disruptive, costly restrictions upon the movement of diseased animals. Consequently, European governments viewed immunization – and, hence, scientific enquiry – as the most promising route to FMD control. Unlike Britain, they had little to lose in terms of exports to FMD-free nations, and with FMD already prevalent, its escape from the laboratory posed less of a threat to agriculture.

Another important factor was that, especially in the veterinary field, Britain had a much weaker tradition of laboratory research than France and Germany. Ever since the mid 19th century, French veterinary students had received a rigorous scientific training that was very different from the practical education imparted by British schools. After graduating, some became laboratory researchers, a career trajectory unknown in Britain. The most famous were Bouley and Chaveau. Based at the Alfort veterinary school, they made discoveries that underpinned Pasteur's later work. This scientific approach to veterinary medicine was reinforced in France and adopted in Germany when Pasteur and Koch made their famous discoveries. Veterinary surgeons in those countries had a far higher status than in Britain. The facilities at their publicly funded schools were far superior to those at cash-strapped British institutions, and their professors engaged in research, whereas in Britain they merely taught. So, while veterinary science took off in France and Germany, Britain boasted only two small laboratories run by Stockman and McFadyean, which as late as 1922 employed just five full-time veterinary surgeons. Small wonder, therefore, that Continental veterinary surgeons favoured a scientific method of FMD control, while their British counterparts preferred a legislative solution.[13]

DOCTORS, VETS AND THE PURPOSE OF SCIENTIFIC ENQUIRY

During the autumn of 1923, in the midst of the second devastating FMD epidemic to hit Britain in three years, questions began to be raised about the continued lack of British FMD research. An increasing number of medical doctors, scientific enthusiasts and critics of the slaughter policy argued that, in the light of recent events, the time had come to extend knowledge of the disease and to search, within the laboratory, for a cheap, humane and scientific control method. However, Stockman's attitude towards FMD research had not changed and he refused to give in to the mounting pressure. He argued that enquiries were unnecessary because veterinary surgeons gained sufficient insights from experiencing FMD in the field and observing experiments that went on 'before our eyes, in nature'. Research, he insisted, was a dangerous enterprise that could contribute little to the control of FMD. In fact, because vaccines typically worked by giving animals a mild case of disease, they could actually encourage FMD spread, and so there was no point in trying to produce them.[14]

Stockman's claims convinced many livestock owners that research posed a threat to their economic interests and he quickly won the support of the empirically oriented veterinary profession. For many doctors, however, his antipathy to scientific enquiry did not make sense. This was because in controlling human infectious diseases they based decisions on very different criteria from those of veterinary surgeons. Slaughter was obviously impossible; therefore, when epidemics struck they aimed not to stamp out disease but to restrict its spread – for example, by removing patients to isolation hospitals (in the case of scarlet fever), vaccinating (in the case of small-pox) or administering antiserum to the diseased (as for diphtheria). The success of this enterprise depended heavily upon the laboratory as it was there that diseases were researched, diagnosed and preventatives prepared. Doctors schooled in the benefits of the laboratory were far more inclined than veterinarians to seek solutions within its walls, and as the 1923–1924 epidemic worsened they became increasingly critical of Stockman's refusal to permit FMD research.[15]

Doctors' opinions received widespread circulation in the medical and public press, and because of the esteem with which the profession was regarded, they had a significant impact upon public opinion. The editor of one of the leading medical journals, the *Lancet*, was especially critical of MAF's 'medieval' attitude to FMD control, its 'devastating want of knowledge' and its 'timidity' in facing up to the small risks involved in experimental enquiry.[16] He argued that circumstantial evidence gained in the field was no substitute for laboratory-based enquiry, which alone could discover definite facts about FMD and devise a vaccine to prevent its spread. The fear of disease escaping from the laboratory was not a sufficiently good reason to ban research, for if such a spirit prevailed within human medicine, 'medical knowledge would stagnate, and the profession would merit the contempt of enlightened men'.[17] The editor of the *British Medical Journal* agreed. He attacked MAF's 'stranglehold on research', and asked how Stockman could possibly offer impartial scientific advice when he was 'so deeply involved in the maintenance of an official policy'.[18] Cheshire doctors caught up in the county's opposition to the slaughter policy (see Chapter 3) similarly emphasized the need for a 'really scientific' method of FMD control and suggested that medical and veterinary professions join forces in the pursuit of this goal.[19] One outspoken medical officer of health, T H Peyton, went so far as to claim that Stockman and his colleagues were 'too conservative and too jealous to learn for themselves, or to allow others to instruct them'.[20]

One of Stockman's more controversial acts was to prevent distinguished medical scientist Professor Beattie from undertaking investigations into FMD. Beattie held the chair of bacteriology at Liverpool University. He was interested in public health, carried out bacteriological investigations on behalf of Liverpool City Council and had been a member of the 1920 FMD Research Committee (FMDRC).[21] When disease broke out in nearby Cheshire he tried to obtain material for experiment; but MAF veterinary staff refused to allow him access to infected animals. In desperation he turned to the Liverpool city meat inspector, who gave him lymph nodes salvaged from cows that had been in contact with infected animals and were probably incubating FMD at the time of slaughter. Beattie fed samples of this tissue to rats, which reportedly went on to develop signs of

FMD. In letters to the *Lancet*, he announced the significance of this finding, suggesting, first, that rats could act as wildlife reservoirs of infection; second, that consuming meat and offal from infected animals could transmit virus; and, third, that rats could be used as experimental animals, thereby reducing the costs and risks involved in large animal research.[22]

Stockman insisted that he had been correct to refuse experimental facilities to men like Beattie. He told the committee of inquiry into the epidemic:

> *Our trouble is, as I have said before, a large number of people who are of no particular account but who wish to be in the limelight want to establish experimental stations all over the country and do some work on foot and mouth. I have held the fort against this single handed up to the present and I have got into rather bad repute – they say, here is a man who is trying to obstruct science instead of helping it. What I say is, with a dangerous disease like this you must investigate it; but you must do so under organized conditions, so that Mr so and so, say, up in Northumberland cannot start a station which is under no supervision. He may let the disease out of that station. We must investigate in certain stations with responsible men who will give an undertaking to the committee that they will carry out the regulations against spread that are laid down.*[23]

Such arguments cut little ice with Walter Morley Fletcher, the extremely influential and outspoken secretary of the MRC. A physiologist by training, Fletcher had worked at St Bartholomew's Hospital, London, and Cambridge University prior to his 1914 appointment.[24] He privately informed medical colleagues that MAF had been 'singularly ill-advised' on FMD:

> *The chief part has been the jealousy of particular leaders of the veterinary profession who, without being at all distinguished in science themselves, have been intensely jealous of the encroachment by human pathologists or medical men upon what they consider their own field.*[25]

Fletcher's claim was not unfounded. For some time, medical scientists had tried to gain access to the field of animal disease. They had justified their interest by arguing that disease involved the same processes in both man and animals and could sometimes be transferred between the two. They pointed to physiologists, who vivisected animals in order to find out how the human body functioned, and to bacteriological and pathological researchers, who used laboratory animals as substitutes for humans. They noted that rabies and anthrax, as investigated by Pasteur and Koch, were diseases of both man and animals; that diphtheria serum was manufactured in horses and small-pox vaccine in calves; and that three Royal Commissions had recently tried to ascertain whether humans were susceptible to the bovine form of tuberculosis (TB). But Stockman and McFadyean shrugged off such connections. They suspected that ambitious medical scientists wanted to take over the field of animal disease research and to turn veterinary medicine into a lesser branch of medicine, stripped of all its authority and claims to expertise. They therefore set out to defend the veterinary field from encroachment. They met with considerable success because, as the leaders of the profession and the heads of its only research laboratories, they exerted considerable influence over the Development Commission, which controlled government funds for animal disease research. During the 1910s and early 1920s, they rejected olive branches proffered by leading medical scientists and obstructed doctors' attempts to initiate enquiries at sites beyond their control, such as university and medical research laboratories. As a result, the field of animal disease research remained extremely small and underfunded.[26]

These machinations infuriated Fletcher. He was utterly convinced of the national benefits of scientific enquiry and felt that scientists worked best when freed from political restraints, a principle that prevailed within his own MRC, which answered not to the Ministry of Health but directly to the Privy Council. Until his death in 1933, he worked hard to impress his vision of research upon the medical profession, and used his influence to encourage specific individuals and branches of research at the expense of others.[27] However, when it came to animal disease research he was powerless to act, and complained bitterly that the failure to make progress was holding back medical advances. He told a friend:

> *I have long been doing everything I can to get proper government support for research into animal diseases and further to link up our work at every point with the study of animal diseases. You may not know, perhaps, of the most incredible obstacles that are put in the way of these two purposes by the present vested interests of the vet world and their ramifications in the Board of Agriculture.*[28]

Fletcher was not the only medical scientist to complain about Stockman and McFadyean's territorial attitude. During June 1920, Sir George Adami, a member of the medical profession and vice-chancellor of Liverpool

University, engaged in a lengthy vitriolic dispute with McFadyean in the correspondence pages of *The Times*. Adami believed that McFadyean's RVC and the Edinburgh and Glasgow veterinary schools should follow the example of the Liverpool school, which had been part of the university since 1904. However, in the field of veterinary education, just as in veterinary research, McFadyean demanded absolute professional autonomy, and he viciously rejected Adami's suggestions.[29] The latter complained: 'Surely a broad-minded man, enthusiastic for the advancement of his science, would welcome and support every movement for increased study and research in his subject', and used the following tale to poke fun at McFadyean:

> *I was taking a Sunday afternoon stroll with my host round his home farm. As we came to the stable yard the great mastiff which had accompanied us made towards the coachman's dog with evident magnanimity. This dog, instead of welcoming him, retreated snarling into his kennel. Whether he thought that the stable yard was his own particular property, or feared that the mastiff would take possession of a kennel several times too small for it, I could not make out; it was the snarl that impressed me.*[30]

The FMD research issue was fought out against this background of inter-professional strife and animosity, and given their previous exchanges, neither side was inclined to mince their words. Stockman derided doctors' 'sentimental' attitude towards the slaughter and their failure to understand the issues involved in animal disease control. He went on to claim that because they had no personal experience of FMD, they had no authority to comment. He told the National Veterinary Medical Association (NVMA):

> *Some of the things that have come from some members of the medical profession in a public position are most inconsiderate and almost insulting. . . I am in no sense closed to good suggestions; but, as often happens, the wrong men run to print when the lime-light is on. . .the majority know too much to speak, the minority know too little to be silent.*[31]

Several leading veterinarians rallied to Stockman's defence. F Hobday, editor of the *Veterinary Journal* and future principal of the RVC, criticized those 'old ladies, clergymen, fossils of various kinds, and even members of the medical profession' who thought they knew the veterinary surgeon's job.[32] The editor of the *Veterinary Record* referred specifically to doctors when he spoke of the 'folly and stupidity of which men may be capable when they write about subjects of which they have little or no understanding',[33] while

J Brittlebank, a member of the NVMA Council, professed his hope that 'the profession will remain loyal to itself. We cannot condemn too severely the interference of outsiders.'[34]

Such comments, however, received little circulation outside the veterinary press. Consequently, veterinarians had little impact on political opinion, which was fast becoming disillusioned both with Stockman's unswerving support for the slaughter policy and his opposition to laboratory enquiry. 'Gravely exercised at the ghastly waste caused by the disease', a Cabinet sub-committee met in December 1923 to consider an alternative strategy of control. Turning their backs on veterinary opinion, members asked Stockman's critic, Fletcher, to recommend an eminent human pathologist to advise them on FMD. He suggested Lieutenant-General Sir William Leishman, a highly respected bacteriologist and director-general of the Army Medical Services.[35]

Fletcher knew that he could rely on Leishman to echo his own views about FMD research, as the two men were good friends. They had worked together in the MRC and on several previous occasions had tried to dislodge Stockman and McFadyean from their positions of power.[36] Stockman, had he known, would undoubtedly have kicked up a fuss; but he was only informed of Leishman's involvement after the invitation had been issued and, according to Fletcher, concurred 'with some reluctance'.[37] Stockman described his feelings later in a letter to Lord Ernle, former president of the Board of Agriculture:

> *I am always willing to consult and collaborate with any good man inside or outside the department; but I think you will realize. . .that it is difficult to sometimes get men who do not understand the problem in all its practical bearings to see that many of their suggestions, when based upon purely abstract principles, may be of very little use; in fact, they may be a waste of time.*[38]

Stockman sent a lengthy letter to Leishman, in which he fully expounded his views on FMD research and control. But he could hardly have hoped to convince that champion of bacteriological research (who gave his name to the tropical disease Leishmaniasis) that laboratory enquiry into FMD was unnecessary. Although, in his report, Leishman acknowledged the validity of Stockman's opinions, he went on to argue that the uncertainty and lack of knowledge that surrounded FMD were seriously impeding efforts to control the disease. He recommended that organized research begin immediately, coordinated by a committee of experts in veterinary science, medicine, bacteriology, immunology and epidemiology. Although MAF should take overall control, it was important for the committee to liaise with MRC experts, and the limited supply of veterinary researchers meant that members of other professions must be directly involved.[39]

Recognizing that laboratory-based enquires were now inevitable, Stockman quickly attempted to ensure that control of this activity did not pass out of veterinary hands. He emphasized that as FMD was an agricultural problem, MAF must take charge of research. He also proposed a predominantly veterinary committee and recommended that work take place at his own and McFadyean's laboratories, and at a new MAF-controlled field station.[40] Fletcher had other ideas. He knew that many doctors would refuse to work 'under' veterinarians, who lay far beneath them on the professional hierarchy. He proposed a more 'balanced' committee of both veterinarians and doctors, and recommended that research also take place at medical research institutes.[41]

Fletcher's vision proved the more influential. In February 1924, Minister of Agriculture, Robert Sanders, appointed an FMD research committee (FMDRC), comprising four veterinary surgeons and six distinguished medical scientists. MAF was to take overall charge of research, which would take place at several institutions, including Stockman's laboratory, the Lister Institute, the MRC's NIMR and a new field station located at Pirbright (see Plate 13). Famous physiologist Charles Sherrington, a former Liverpool University professor and an MRC colleague of Fletcher's, became the first chairman of the FMDRC. He soon resigned due to ill health, and Leishman took over until June 1926, when he died on the same day as Stockman.[42]

It is difficult to know how McFadyean and Stockman judged the outcome of this controversy. As we have seen, the FMD research question reopened long-standing feuds over the authority, territory and appropriate activities of the healing professions. It illustrated important differences in their attitudes towards disease research and control, and exposed the profound gulf that then existed between the established, authoritative medical profession and the struggling veterinary trade. So, were Stockman and McFadyean outmanoeuvred and overthrown by influential members of the medical profession? Or, in the end, had they lost the battle but won the war? After all, Stockman was not only engaged in a controversy over animal disease research; he was also fighting to defend the unpopular slaughter policy (see Chapter 3). His critics viewed the government's decision to initiate FMD research as an important concession, and, in the belief that alternative control methods would soon be forthcoming, they allowed Stockman to continue with the slaughter. This was not his only victory. As the next section will reveal, he and his colleagues went on to achieve a substantial degree of control over the organization and direction of FMD research, with important consequences for the nation.

In another sense, however, the commencement of FMD research in Britain marked the end of an era. Before long, Stockman was dead and McFadyean retired. Their 20-year period of professional dominance was at a close and the ethos of 'splendid veterinary isolation' died with them. Other new research initiatives were already beginning. In 1921, a group of Scottish farmers funded the formation of the Animal Diseases Research Association. This body employed veterinarians to investigate farming problems and founded the Moredun Research Institute near Edinburgh, which still exists today. In 1922, research into canine distemper began at the MRC-controlled NIMR, funded by subscribers to the *Field* magazine. The following year, the Development Commission endowed a professorship and erected a new Institute of Comparative Pathology at Cambridge University.[43] Together with the formation of the FMDRC, these activities extended the veterinary research field and permitted closer collaboration between veterinary, medical and agricultural scientists. Meanwhile, a younger generation of veterinarians rose to prominence and set about shaping the veterinary profession of the future.

KEEPING FMD OUT, 1924–1938

From its formation in 1924 until the outbreak of World War II, members of the FMDRC met every one or two months. They discussed administration and staffing, reviewed scientists' written reports of research progress and decided upon subjects for future investigation.[44] Although it was nominally an independent body, MAF exerted substantial control over its activities. Officials requested certain experiments, arranged the publication of research results and helped to select committee members and scientists, while ensuring that most of the latter were veterinary surgeons. MAF also stationed a veterinary inspector at the Pirbright field station to check that staff took the necessary precautions against disease spread. Following Stockman's early death, his responsibilities were divided so that Ralph Jackson became CVO, while W H Andrews was appointed head of veterinary research. Both gained permanent places on the FMDRC. Like Stockman, they feared the escape of virus from the laboratory and insisted that the committee turn down all independent requests to conduct FMD research.[45] The only exception to this rule was Professor Beattie who, probably as a result of political pressure, was permitted to carry out unpaid experiments at Liverpool University. This arrangement lasted for only two years. Hindered by his other responsibilities and demoralized by the committee's lack of regard for his efforts, Beattie abandoned the field in 1926.[46]

From 1931, FMD research became subject to the scrutiny of the Agricultural Research Council (ARC), a new organization established against MAF's wishes with the object of guiding, and providing greater freedom for, agricultural research. In 1937, MAF successfully resisted moves to place the FMDRC under the ARC by claiming that only MAF was capable of preventing the FMD virus from escaping from the laboratory.[47] It retained this monopoly of power until the 1950s, when reluctantly, and in piecemeal fashion, it handed over to the ARC.

Until that time, MAF's notions of how FMD ought to be controlled exerted a powerful influence over scientists' activities. The FMDRC's remit was 'to direct and conduct investigations into FMD, either in this country or elsewhere, with the view of discovering means whereby the invasions of disease may be rendered less harmful to agriculture'.[48] To most members of the British public, making FMD 'less harmful' meant searching for a vaccine or serum with which to prevent disease spread. To MAF, however, the best way to make FMD 'less harmful' was to stop it from entering Britain in the first place because the controversial slaughter policy would not then be required. Under its direction, British FMD scientists devoted most of their time between 1924 and the outbreak of World War II in investigating where FMD had come from and how it was spreading, a somewhat one-sided approach. By contrast, in European countries where FMD was endemic and foreign invasions of virus less significant than internal disease dissemination, making FMD 'less harmful' meant limiting its spread and reducing its impact upon meat and milk production. Consequently, Continental scientists focused far more upon serum production and vaccine research than their British counterparts.

Before they could begin work, British FMD scientists had to familiarize themselves with the techniques and findings of a quarter of a century of European FMD research.[49] For example, German scientist Professor Waldmann had recently discovered that guinea pigs could be used as experimental animals. Also, along with French Professors Vallée and Carré, he had demonstrated the existence of three types of FMD virus, O, A and C, and had shown that infection with one type did not produce immunity against the others.[50]

British scientists devoted much effort to discovering the virus type responsible for British FMD outbreaks. Their results helped to determine whether outbreaks were connected, and because the distribution of virus types varied around the world, they gave some indication of the country of origin. Researchers also assessed whether wildlife was susceptible to infection, under what conditions virus survived in the environment, and how long it could live on imported substances such as dried milk, meat and packing materials.[51] MAF used their findings as a basis for new legislative orders, such as the Boiling of Animal Foodstuffs and the Importation of Meat (Wrapping Materials) Orders, which aimed to prevent FMD-contaminated imports from coming into contact with susceptible animals.[52]

Another important line of work was serum testing, and between 1930 and 1934, CVO John Kelland trialled this product in the field. But while, on the Continent, serum was employed as a first line of defence against FMD spread, Kelland viewed it rather differently, as 'an additional weapon in the armoury of the state's warfare'. It was not intended to replace the slaughter policy, but to protect at-risk animals located near infected farms.[53] It could also reduce FMD importation into Britain, as 'the disease undoubtedly comes here from abroad, and if the committee's work were successful and results could be put into operation abroad, this country would reap the benefit'.[54] Unfortunately, field trials showed that serum failed to produce reliable, long-lived immunity against FMD.[55] This was well known on the continent but did not matter, as serum still reduced the clinical severity and economic cost of FMD. But in Britain, where the slaughter policy made the long-term clinical effects of FMD irrelevant, the main problem facing MAF was how to prevent disease from spreading. Serum use actually contributed to that problem: in making symptoms less obvious, it hindered the early diagnosis and control of FMD.[56]

Because MAF officials were convinced that slaughter was and would remain the best method of controlling FMD in Britain, they saw little reason to encourage vaccine research. In any case, this was a technically difficult line of enquiry: virus culture still posed problems; early experiments upon vaccines devised on the Continent met with little success; and new complexities emerged during the 1920s and 1930s, when scientists discovered that each virus type consisted of a number of strains that varied in their propensity to infect different species and in the clinical signs that they produced.[57] In 1937, an ARC sub-committee reviewed the FMDRC's progress and suggested that it should pay more attention to vaccine research. However, it was well aware of the difficulties involved, and viewed vaccines in the same limited way that Kelland had viewed serum: they were to supplement, not replace, the slaughter policy and might be useful overseas to prevent the export of virus to Britain.[58] Scientist J T Edwards rose to the challenge, but succeeded only in devising extremely complex vaccination regimes that would have been quite impossible to apply in the field.[59]

MAF officials knew full well that many members of the public believed scientists were primarily engaged in discovering a vaccine that would one day replace the slaughter policy. To dampen down such expectations they prevented scientists from publishing their findings freely in scientific journals, a restriction that one researcher, H Skinner, viewed 'as necessary. . .to avoid false interpretation by those who lobbied against the slaughter policy'. Results were, instead, compiled into occasional FMDRC progress reports, released in 1925, 1927, 1928, 1931 and 1937. These official publications repeatedly emphasized the many difficulties involved in vaccine research and the poor prospects of success. Skinner later elaborated:

> *The word vaccine, you had to be very careful not to use it too much because the fear was that if the public knew there was a vaccine available there'd be a clamour for abandoning the stamping out policy, and the Ministry really wouldn't stand for that. They knew that stamping out was the only thing, to stamp it out. Couldn't have people vaccinating animals against it.*[60]

It is clear, therefore, that for the first 14 years of the FMDRC's existence, British FMD research followed a distinctive path that was selected in accordance with, and helped to shore up MAF's existing FMD control policy. The official conviction that national freedom from FMD had to be maintained at all costs, and that import barriers represented the first line of defence against FMD and slaughter the second line of defence, all had a direct impact upon scientists' activities. Under MAF's direction they looked not for new methods of FMD control, but for ways of improving the existing control policy and thereby reducing the criticisms that had surrounded it since its late 19th-century inception. However, one only has to look to the Continent to see that British FMD research could have taken a very different direction.

A WARTIME THREAT

The late 1930s saw a dramatic shift in MAF's approach to FMD research, as a result of which British FMD scientists began, for the first time, to pursue similar goals to their Continental counterparts. Directing this change was CVO Daniel Cabot. He evaluated the methods available for FMD control and decided that, in future, it might prove difficult to prevent FMD invasion. He also acknowledged that under certain conditions, slaughter might fail and would have to be replaced with immunization. In anticipation of this event, he arranged for the FMDRC to stop epidemiological investigations, resume serum trials and prioritize vaccine research.

Driving Cabot's rethink was the fear that Britain could fall victim to germ warfare. He was extremely concerned that under wartime conditions, hostile countries would manage to introduce the FMD virus into Britain in such quantities that a devastating epidemic would result. Under such circumstances, the slaughter policy would probably fail. Within no time at all, FMD would become endemic and British meat and milk production would fall. This was far more than just an economic issue. During wartime, it would be impossible to import replacement supplies, and the resulting drop in domestic consumption would seriously impact upon the health, morale and fighting ability of the British people.

The likelihood of Germany employing germ warfare was first considered by the government's Committee of Imperial Defence in 1936 as part of the contingency planning for war. It concluded that Germany probably did not possess human biological weapons and was unlikely to develop any because it was extremely difficult to produce and distribute them without risk to German civilians. FMD, however, was a different matter. Because the disease was endemic in Germany, most livestock possessed some resistance to infection, and so scientists could develop weapons without fear of initiating an epidemic. Germs could be easily distributed by air; they were highly contagious and could spread rapidly. Moreover, all British livestock were susceptible to infection.[61]

Initially, Cabot was not convinced by the committee's conclusions. After all, there were three types of FMD virus, and infection with one did not produce immunity to the others. German livestock were unlikely to have solid resistance to all three, and the highly contagious FMD virus could easily escape from the research laboratory to infect them.[62] But in June 1938, the situation changed dramatically, as Professor Waldmann, head of Germany's FMD research institute on the Island of Riems, announced that he had discovered the first effective FMD vaccine.[63] Unlike serum, vaccines provided long-lasting resistance to infection and were a far more effective means of limiting FMD spread. The implications of this discovery were potentially devastating. German scientists could now devise a biological weapon without risk to their own livestock and without fear of reprisals.[64]

Cabot quickly realized that if Britain was subject to germ warfare, there was only one way of limiting the damage to the nation's meat and milk supply: immunization. There were two options open to him. He could either defend Britain's livestock using Waldmann's new vaccine, or he could resort to serum treatment. He turned to the FMDRC for advice. Members had been in communication with Waldmann and were not overly impressed with his claims. It appeared that that his vaccine was not fully effective against all three types of virus and its method of production was extremely complicated. By contrast, serum was cheap and easy to prepare, and if gathered at the site of an outbreak, would provide adequate though short-lived protection against the prevailing virus type. Cabot therefore decided that serum should provide the nation's main defence against biological attack.[65]

There was a snag, however. British scientists had no idea how to produce serum, having bought supplies for earlier experiments directly from Waldmann's laboratory. The FMDRC hurriedly dispatched scientist H Skinner on a fact-finding tour of the Continent. He returned ahead of schedule when war broke out, clutching plans of the Danish government's serum-production apparatus. MAF installed a carbon copy at Pirbright, where all FMD scientists were now based, under the control of the director, Ian Galloway. During the years that followed, they produced thousands of gallons of serum from the blood of artificially infected animals and transported it by car to requisitioned cold stores in Surrey. They used the same apparatus to produce cattle plague serum for fear that Germany would also use this deadly disease as a weapon. Meanwhile, MAF distributed portable serum-production equipment to ten centres throughout the nation so that, if necessary, it could be collected from local animals that had recovered from FMD.[66]

Cabot did not lose sight of the fact that in the long term, subject to various technical improvements, vaccination offered a more reliable and longer-lasting method of protection. He therefore directed those scientists who were not engaged in serum production to stop epidemiological enquiries (which, in any case, were hindered by the wartime lack of guinea pigs) and focus their attention upon vaccine research.[67] This meant that for the first time, British FMD scientists were pursuing similar goals and lines of enquiry as their Continental counterparts. But this shift made sense because if Britain did fall victim to biological attack, it would have experienced FMD in the same manner as Continental nations and had recourse to the same methods of control.

During the war, all research publication was suspended and regular FMDRC meetings ceased. A core committee continued to gather until 1944, when – upon the death of J Arkwright, chairman since 1931 – Cabot assumed direct control. Several scientists left to join the forces or work elsewhere; but two extremely capable veterinary surgeons, J Brooksby and W Henderson, remained at Pirbright and carried out much valuable work.[68] Meanwhile, at the biology department at Porton Down, Wiltshire, a much larger body of scientists was carrying out top-secret investigations into the offensive and defensive possibilities of other types of germ warfare. In collaboration with the US and Canada, they developed anthrax-filled cattle cakes, anti-livestock weapons intended for retaliatory use only; 5 million lay stockpiled by the end of the war. Scientists also devised anthrax bombs and carried out open-air tests of their weapons on Gruinard Island, Scotland, which left the island completely uninhabitable.[69]

In the event, Germany did not use biological weapons. The British government's elaborate plans to defend the nation were not put into practice and MAF maintained its traditional FMD control policy throughout the war. Outbreaks occurred frequently, mainly because of a rise in pig-keeping, which MAF had encouraged since pigs could eat kitchen waste and supplement wartime food rations. From time to time, scraps of foreign infected meat found their way into un-boiled pigswill and caused FMD.[70] The years 1941–1942 saw the largest epidemic since 1922–24 and drew complaints that the slaughter policy wasted valuable meat. Cabot responded by allowing butchers to salvage the carcasses of healthy animals that were culled alongside their diseased contacts, a practice that had been stopped in 1923 because it slowed the elimination of disease.[71] He also took advant-age of the British government's extensive wartime control over the nation's food supply and diverted supplies of suspect South American meat away from rural areas and towards the cities, where the government's Waste Food Board ran swill-boiling plants. With the exception of a brief period around D-day, this scheme ran from 1942 until 1954, when the government relin-quished control over the food industry, and MAF and the Ministry of Food joined to form the Ministry of Agriculture, Fisheries and Food (MAFF). MAF thought the scheme effective in reducing FMD incidence, but did not inform the general public for fear of complaints from urban residents, who might prefer home-killed meat but received only imported produce.[72]

It is debatable whether Germany ever intended to attack Britain using the FMD virus. Certainly, Waldmann made few attempts to keep vaccine discovery secret, and readily shared details of production techniques with his British counterparts. Throughout the war, British intelligence kept a close eye on activities at his laboratory and interrogated several German scientists about their country's progress in biological weapons development. Reassuring facts were discovered: Waldmann's research institute on the Island of Riems was apparently incapable of producing large quantities of FMD vaccine and did not hold large stocks; moreover, Hitler was not interested in biological warfare.[73] However, a recent account in *The Times*, based upon German sources, suggested that the German Secret Service was, in fact, planning an attack. It reportedly organized large-scale virus production at Riems, carried out successful weapons-testing upon a Russian island and, by the end of the war, was capable of attacking Britain.[74]

THE COLD WAR AND BIOLOGICAL WEAPONS RESEARCH

British fears of germ warfare did not cease with Germany's defeat. The descending 'Iron Curtain' around Soviet-controlled countries in 1947, the 1950 Korean War and the signing of the Warsaw pact by Eastern bloc

European nations in 1955 all marked the rise of a new and hostile power that was allegedly capable of employing biological weapons against the West. Believing that the nation should be ready to defend itself against attack and to retaliate with similar weapons, the post-war British government allocated large resources to both defensive and offensive biological weapons research. During the late 1940s, the Defence Research Policy Committee of the Ministry of Defence (MOD) recommended a massive increase in effort to ensure that, by 1957, biological weapons would be of comparable readiness to the atom bomb. To achieve this goal, Britain entered into tripartite research agreements with Canada and the US. As a general rule, the US – which unlike Britain had not signed the 1925 Geneva Protocol prohibiting the use of offensive biological weapons – carried out short-term weapons research, while British scientists, mainly based at Porton Down, focused upon more fundamental problems.[75]

FMD was awarded high priority within this research programme. Canada and the US felt vulnerable to attack on account of their national freedom from FMD and susceptible livestock population. Their fears increased when, in 1952, Canada suffered its first ever FMD epidemic, later attributed to a German immigrant farm worker whose clothes were contaminated with the virus. In planning their defences, all three nations decided that they should possess sufficient supplies of vaccine and sera, facilities for their production at short notice, and the intelligence to provide an early warning system against attack. They also believed that they should be in a position to retaliate if necessary.[76]

As already described, the devastating FMD epidemic of 1923–1924 caused the government to override MAF's resistance to FMD research. Such pressures had never arisen in Canada and the US, where FMD incidence was far lower in Britain. Consequently, fears that virus would escape from the laboratory prevailed and no research took place until 1954, when the US opened its first FMD research laboratory on Plum Island, off the coast of Long Island. Prior to that date, all three nations depended upon work taking place at Pirbright. The MOD regarded this situation as a prime opportunity to improve its relationship with the US.[77] In August 1951, it recommended that a recently proposed seven-year plan to extend laboratory accommodation, facilities and research at Pirbright should, for maximum military value, be intensified and completed in two years.[78] Scientists would then be in a position to undertake a 'study of offensive possibilities and defence requirements in their widest sense'.[79]

Because MAF was responsible for maintaining disease security at Pirbright, selected officials became privy to the MOD's plans.[80] CVO Thomas Dalling reportedly told his staff that the goals of research were twofold: to protect livestock against enemy attacks (possibly by large-scale immunization, although this method was not to replace slaughter during 'peacetime') and to enable Britain to take 'offensive actions versus a possible enemy'.[81] However, many members of the Pirbright Institute's governing body (which replaced the FMDRC after the war) were not informed, nor was the scientific community at large.[82] This level of secrecy placed MAF officials in an awkward position. How were they supposed to explain a government grant of UK£250,000 to Pirbright when post-war financial stringency meant that the total sum available for 12 other agricultural research institutes during 1952–1953 was only UK£300,000? The Treasury suggested that they pass off the money as 'an acceleration of research of purely civilian interest', a somewhat implausible explanation. Sir Harold Parker at the MOD had a different idea:

> *The full reasons for the expansion and acceleration certainly cannot be given. I would not, however, see any difficulty in your saying. . .that there is considerable defence interest in this work from the point of view that the ensuring of our food supplies in war is vital, and that a solution of the FMD problem is a very important factor.*[83]

In an obvious reference to biological warfare research, Sir Donald Vandepeer, MAF permanent secretary, replied, 'We should certainly not have placed it [FMD research] at all high on our list of priorities for research from the point of view of ensuring future food supplies if it had not been for the defence considerations, which are just the sort of thing you would not want us to mention.'[84] In the end, officials decided, under pledge of secrecy, to tell a few concerned parties of the MOD's involvement.[85]

National economic difficulties caused the revaluation of many defence plans in May 1952. However, the MOD continued to prioritize FMD research and demanded the completion of building work at Pirbright by the end of the year. In the event, the project ran five years behind schedule, its progress hindered by clashes between Pirbright's director, Ian Galloway, members of the governing body, the Agricultural Research Council (which, in 1951, took charge of research), and MAF (which retained control over financing, staffing and disease security).[86] Ernest Gowers, chairman of the 1952 committee of inquiry into FMD (see Chapter 6), summed up the situation in a letter to G H R Nugent, MAF parliamentary secretary:

> *You ought to look at the Pirbright problem personally. The scientists there are splendid people doing splendid work. But the system of control is just chaos. The governing body say they are responsible to the Minister* [of Agriculture]; *but the ARC say*

> *they are responsible to them. Whichever of the two is supposed to control the governing body, the governing body itself is quite incapable of controlling Galloway,* [who has] *been charged (so he says) with highly secret work which he cannot possibly tell his governing body about because they have not been passed by MI5!*[87]

By the time building work was completed at Pirbright, the fear of biological attack and the status of biological warfare research in Britain had fallen considerably, partly because the MOD regarded the possession of nuclear arms as an adequate deterrent to biological attack, and also because the US had adopted an increasing share of research.[88] The fate of the FMD biological weapons programme remains a mystery, although more information may well come to light when remaining government files are declassified.[89] Clearly, however, the MOD's generous investment was of great benefit to Pirbright. By 1958, 29 full-time research staff were employed at the laboratory, compared to just five in 1939. They worked not only in the traditional fields of vaccine production and serology, but also in new and rapidly expanding areas, such as genetics, biophysics and biochemistry.[90]

As the threat of biological warfare waned, MAF felt it less and less likely that FMD vaccines would be needed in Britain. Officials regained confidence in the traditional FMD control policy, and the problem of keeping FMD out of Britain returned to the top of their agenda. But the return to a pre-World War II approach to FMD control was not matched by a similar shift in research activity. Contrary to what one might expect, MAF officials did not, in the aftermath of the biological warfare threat, view vaccine research as unnecessary and irrelevant. Instead, they decided that vaccination – if used overseas – was a prime means of preventing FMD importation. They therefore encouraged scientists to intensify their efforts in this field.

MAF stuck with vaccination partly because of recent advances in technology, which considerably enhanced the prospects of success. Another motivating factor was a devastating FMD epidemic (see Chapter 6) that swept across Europe and into Britain during 1951–1952. In its aftermath, CVO Thomas Dalling (who left his post in 1952 to join the United Nations Food and Agriculture Organization – the FAO)[91] and his MAF colleagues decided that only a systematic international attack upon FMD could possibly protect Britain from future invasions. To this end, they proposed that the FAO establish a European Commission for the Control of FMD (EUFMD), and laid down a FMD control scheme to which member nations should adhere. The ultimate goal of the scheme was FMD eradication. It was to be achieved by livestock movement restrictions, plus one or more of the following: slaughter; slaughter and vaccination; vaccination to maintain a totally immune animal population; or vaccination in barrier zones along frontiers. In December 1952, the FAO considered and approved these plans and in July 1954, after just six nations had joined, the EUFMD held its first meeting. More nations joined in subsequent years.[92]

In helping to set up this new organization, Dalling managed to institutionalize, at an international level, MAF's vision of FMD control and its ideal of national 'FMD-freedom'. The four-tier system adopted by the EUFMD was designed to engineer consensus between nations that experienced and responded to FMD in very different ways. Initially, it permitted and even encouraged vaccine use; but as disease incidence dropped, it directed nations along the road towards a fully fledged British-style slaughter policy. The success of this venture depended upon the production of large quantities of safe, potent vaccines, and that in turn required substantial research efforts. MAF wanted Pirbright to play a leading role, and was extremely pleased when, in 1957, it became the FAO-designated 'World FMD Reference Laboratory'.[93] Five years later, MAF and ARC handed responsibility for vaccine work to the pharmaceutical company Burroughs Wellcome and Co, which established a laboratory within the confines of the Pirbright field station.[94] As a result of these developments, the declining biological warfare threat failed to impact upon British FMD vaccine research, which continued in earnest, having found a new purpose and a new institutional home.

This history shows clearly that for much of the late 19th and 20th centuries, the goals of FMD research and control were inextricably linked, and that for political, geographical, professional, commercial and scientific reasons, they varied between nations and over time. Research did not so much shape FMD control policy as grow out of it. Indeed, British FMD scientists' activities only make sense when viewed in the light of MAF's long-standing drive for national FMD-freedom, and its mid-century fear of germ warfare. Had MAF failed to gain and retain control of FMD research, and had Britain not experienced the crises of World War II and the Cold War, then research might have followed a very different path and had a very different relationship with disease control policy. But as Chapter 6 explains, MAF's vision of FMD research was not universally accepted, and during the early 1950s, it became the subject of a wide-ranging controversy.

Chapter 6

The 1951–1952 Vaccination Controversy

FMD RETURNS

Towards the end of April 1952, Rupert Guinness, second Earl of Iveagh, chairman of the famous brewing company and elder brother of the late Walter Guinness, former minister of agriculture, surveyed his Elveden estate. Just seven years earlier, his herd of pedigree Guernsey cows had contracted foot and mouth disease (FMD) and been slaughtered. Now their replacements had suffered the same fate. Iveagh was angry, upset and his patience was wearing thin. Why, he asked, hadn't the Ministry of Agriculture and Fisheries (MAF) devised a better way of protecting Britain's herds against FMD? Were officials not aware that in Denmark, The Netherlands and Germany livestock were now vaccinated against the disease? Surely it was time to reassess the situation, as in the current scientifically advanced age, vaccination was clearly preferable to the outdated Victorian policy of slaughter.[1]

Iveagh was not alone in thinking that the slaughter policy had had its day. By the spring of 1952, Britain was in the grip of an FMD epidemic, which had begun the November before when birds migrating from FMD-infected European nations had allegedly carried virus into coastal parts of eastern and southern England. As the disease spread, seemingly unchecked by the slaughter policy, into the Midlands, Scotland and Wales, criticisms of MAF's actions and calls for FMD vaccination became ever louder.[2] Stung by public opinion, MAF and its supporters responded with a vigorous defence of the status quo. The resulting controversy was played out in Parliament, the national press and in private hearings of the Gowers Committee of Inquiry into the epidemic. Participants included British and foreign scientists, veterinarians, selected livestock owners, journalists, MAF officials and members of the Agricultural Research Council (ARC). The dispute was never fully resolved, and died away only when disease incidence fell.

The 1951–1952 row over vaccination was the first major challenge to MAF's FMD control policy for almost 30 years. The vigorous criticisms of slaughter that had characterized the 1922–1924 epidemics (see Chapter 3) – faded in the spring of 1924 and remained muted during the years that followed. This was partly because both the 1922 and 1924 committees of inquiry into FMD backed slaughter and dismissed calls for a return to the earlier policy of isolation. Furthermore, during the years that followed, slaughter was more successful in limiting FMD spread. In addition, MAF and its many supporters acted to exclude critics from the processes of policy-making and execution. Legislative changes, introduced in accordance with the committees' recommendations, resulted in a more centralized FMD control policy and enhanced MAF's powers at the expense of the local authorities, many of which had objected to its handling of the 1922–1924 epidemics.[3] Meanwhile, that avowed advocate of slaughter, the National Farmers' Union (NFU), consolidated its hold over farming politics and eliminated dissenting voices from policy discussions. The 1926 discovery that imported meat could convey the FMD virus further diverted farming attention away from the slaughter policy, and as agriculture became more prosperous during and after World War II, farmers became less inclined to complain about the costs of controlling FMD.[4]

This ceasefire ended with the 1951–1952 FMD epidemic. By 20th-century British standards, this was a moderate sized epidemic: 600 FMD outbreaks occurred during a 12-month period and led to the slaughter of 85,000 livestock at a cost of UK £3 million in compensation.[5] When compared to earlier and larger epidemics – including that of 1941–1942, which barely drew mention in the newspapers – the outcry that accompanied it was out of all proportion to its size. And although the controversy featured traditional farming complaints about the social, psychological and economic hardships of FMD control, it was also fuelled by a novel set of criticisms that, in their source, scope and mode of expression, differed significantly from those that had been voiced before.

Britain, after World War II, had an economy bordering on bankruptcy and an empire on the brink of collapse, and its status as a world power was in terminal decline. Fearing an economic boom and slump similar to that which had followed World War I, the government maintained many wartime controls over production and supply. Consequently, the austerity that had characterized the war years continued well into the 1950s. In an attempt to reduce spending on imports and to meet the threat of a world food shortage, the government passed the 1947 Agriculture Act. Which promoted and sustained 'a stable and efficient agricultural industry', this ushered in an era of increasingly intensive farming, and brought prosperity to the many participants.[6]

The economic crisis led to new protests over the cost of FMD control. An additional concern was whether the rise of motor transport and the shift to larger farms with higher stocking densities had enabled FMD spread to outstrip the capacity of the slaughter policy.[7] There were further complaints that controlling FMD by slaughter wasted meat, a highly valued foodstuff that was still rationed.[8] The *Daily Telegraph* protested:

> *This is the position after the disease has been notifiable for 83 years, after the slaughter policy has been pursued for 60 years, and after intensive research has been pursued for 30 years. What was once an inconvenience threatens to become a disaster at a time when we cannot afford to lose a morsel of meat.*[9]

But while these issues were undoubtedly important, the most important cause of the 1952 backlash against the slaughter policy was the news that European (and South American) nations had recently adopted a new and reportedly successful method of FMD control – vaccination.

The British 1951–1952 epidemic was an offshoot of a pan-European epidemic, which comprised over 900,000 outbreaks.[10] As the disease extended across the continent, alarmed governments and livestock owners reached for FMD vaccines, using them alongside the more traditional method of serum administration. Although vaccines were not yet produced in sufficient quantities to allow systematic, Europe-wide application, recent technical advances had increased their availability to a level at which strategic use became possible.[11] France, Belgium, Holland, Denmark, Switzerland and Sweden tried to create new barriers to FMD invasion by immunizing susceptible livestock near country borders; in Sweden and Switzerland, 'ring vaccination' took hold, whereby animals in the region surrounding a disease outbreak were immunized. Many nations chose to selectively immunize vulnerable or valuable stock, and they also used vaccines to dampen down the volume of infection within a given geographical area. In some regions, farmers were allowed to vaccinate voluntarily at their own expense, while elsewhere, vaccination was enforced by the state.[12]

Vaccination did not prevent FMD from extending across Europe, partly because vaccines were in short supply, and also because a new type of virus emerged mid epidemic, which had not been included in the vaccine.[13] Nevertheless, many European scientists, livestock owners and government officials believed that vaccine use had been beneficial in limiting the size and economic impact of the epidemic. They also felt that it had helped farmers psychologically by allowing them to take personal responsibility for the health of their stock and abolishing much of the uncertainty about whether disease would strike.[14]

As FMD spread throughout Britain, journalists began to report upon the purportedly successful use of vaccination on the Continent.[15] This news had extremely important implications. As Chapter 5 clearly showed, MAF habitually withheld information relating to vaccine development from the public domain. But during 1951–1952, its policy of secrecy became unsustainable as newspaper readers learned – many for the first time – that FMD vaccines were available and could help to control the disease. The previously marginalized opponents of the slaughter policy gained a new weapon of attack, and as MAF had long feared, they began to clamour for FMD vaccination in Britain.

CALLS TO VACCINATE

British proponents of FMD vaccination were selected veterinary surgeons, journalists, and farmers. Some had always objected to the slaughter policy; others had only tolerated it in the (somewhat mistaken) belief that British scientists were striving to discover a vaccine with which to replace it.[16] They viewed slaughter as a 'barbaric and medieval' policy, 'a deplorable and abject confession of defeat. It is Medicine Man stuff, a survival from the unscientific past.'[17] Vaccines, on the other hand, appeared modern, scientific and humane. They felt that under a vaccination policy, farmers would cease to suffer the periodic loss of thousands of animals, the cost of FMD control would diminish, and there would be no further loss of meat. For influential upper-class livestock breeders such as Lord Iveagh, whose valuable pedigree animals had once been exempt from the slaughter (see Chapter 3), vaccine discovery also raised the prospect of wresting control of FMD from an undiscerning centralized government department and its 'one-size-fits-all' policy.[18]

Adding fuel to the flames of this controversy was Dr William Crofton, a maverick Irishman who believed that he personally held the secrets to FMD vaccine production. Crofton had been known to MAF since the later 1930s, when, in a stream of letters to ministers, officials and the press, he claimed to have disproved the generally accepted notion that viruses caused disease. In his booklet *The True Nature of Viruses*,[19] he contended that viruses were but one stage in the bacterial life cycle, and claimed to have cultured the bacterial agent of FMD and used it to produce an effective vaccine. Although dismissed as a quack by the medical and veterinary establishments, Crofton was no swindler. He seriously believed that his discovery was of great benefit to the nation, and he managed to convince many well-connected politicians and aristocrats of his good intentions. These influential individuals wrote countless testimonials on his behalf to the minister of agriculture, and the resulting political pressure forced reluctant officials to commission a scientific test.[20] As they had anticipated, FMD researchers at Pirbright proclaimed the vaccine worthless. But Crofton declared the test invalid because it had been carried out on guinea pigs, not farm animals, and made an unsuccessful bid for further investigations.

At the outbreak of World War II, Crofton offered his discovery to MAF, free of charge, for the duration of the war. Rebuffed, he headed to South America in order to carry out further tests and to market his vaccine. He did not give up on MAF, however. To the evident irritation of officials, he continued to press his claims, both in person and indirectly via a collection of worthy and notable patrons.[21] As the years passed, he grew increasingly bitter about his failure to convince any member of the scientific establishment, at home or abroad, to take up his FMD vaccine, and his communications with MAF became ever more hostile. In spring 1952, he added his voice to the rumblings against the slaughter policy and cast scorn on government-backed FMD scientists, who had seemingly failed to discover a vaccine in spite of high-level funding.[22] His claims stuck a chord with many pro-vaccinators, who regarded him as a beleaguered hero who had for many years engaged in a valiant, altruistic battle against a politically motivated, intransigent establishment. The editor of the *Daily Telegraph* argued:

> *Surely any method offering* prime facie *possibilities of this kind deserves renewed investigation, under conditions agreed on all hands to be satisfactory, by the highly qualified scientific organization which the government has at its command. It must never be said that, in such an emergency as the present, claims which might be valid should be rejected without full and fair trials.*[23]

Crofton's criticisms helped to focus public attention upon the organization and achievements of British FMD research. It was commonly assumed that the purpose of FMD research was to find a vaccine with which to replace the slaughter policy. Consequently, the adoption of vaccination on the Continent and MAF's continuing adherence to slaughter caused many commentators to conclude that British scientists lagged far behind their Continental counterparts. They tried to rationalize this state of affairs. Were British scientists competent to undertake vaccine research? Did they lack funds? Had they taken sufficient notice of discoveries made overseas? Alternatively, was MAF to blame? Had it misdirected British FMD research, or failed to adjust its FMD control policy in accordance with the latest scientific advances? Whatever the reason, commentators agreed that MAF ought to report – for the first time since 1937 – upon scientists' progress. It should also try to reassure the public that it was aware of the need for better FMD vaccines and was encouraging research that would eventually allow the discontinuation of the slaughter policy. As the *Daily Telegraph* warned: 'if farmers are convinced that strenuous and objective efforts are being made to find other less wasteful ways of combating the disease, then and then alone will they continue their reluctant compliance [with the slaughter policy].'[24]

THE MINISTRY STANDS FIRM

These criticisms stirred MAF officials into action while their belief that slaughter was by far the best policy still garnered substantial support, they realized that farmers opposed to it could easily obstruct its workings. For example, they could neglect to report early disease symptoms, refuse to keep away from infected premises, or mix with individuals who had been in contact with diseased animals. Such behaviour would facilitate the spread of disease, and further criticisms of the slaughter and renewed calls for vaccination would inevitably follow. It was, therefore, in MAF's interest to win over farming opinion (if not public opinion, in general) by justifying its opposition to FMD vaccination.

As Chapter 5 showed, vaccine development had headed the list of British FMD research priorities ever since the outbreak of World War II, and by 1952, scientists had made considerable progress. They had devised important techniques for testing vaccines, made considerable advances in vaccine potency and safety, and, between 1946 and 1951, assisted in the control by vaccination of a severe epidemic in Mexico.[25] But MAF officials never intended to use vaccines in Britain except in response to germ warfare. They never publicly communicated this fact, partly because of the secrecy surrounding biological weapons and partly because they feared that the public would simply refuse to accept that slaughter was the last word on FMD control. During the early 1950s, their strategy of concealment became untenable, thanks to widespread press reports that made vaccine discovery common knowledge. Consequently, officials had to find new ways of maintaining public confidence in the traditional FMD control policy and in British FMD research.

Denying that vaccines existed was simply not an option, as this would confirm public suspicions that British FMD research was misdirected or inadequate; but admitting to their possession would provoke additional demands for a change in FMD control policy. Officials therefore argued – in Parliament and in the press – that although vaccines existed, slaughter was, for scientific and economic reasons, still the best policy. Their claims were widely supported by members of ARC (which had, in May 1951, taken control of the Pirbright research programme),[26] scientists connected with FMD research, and veterinary and farming leaders, several of whom wrote press articles that backed up MAF's stance.[27]

MAF spokesmen and their supporters used several key arguments in support of their belief that slaughter was the best FMD control policy. First, they declared vaccination more expensive than slaughter. Their calculations showed that to maintain complete national immunity to infection, vets would have to inoculate all susceptible animals against all three strains of FMD virus every four months, at an annual cost of UK£13 million. By contrast, over the last 25 years, the slaughter policy had cost an average UK£176,000 a year in compensation. Second, they argued that vaccines were dangerous because animals did not develop FMD immunity until 14 days after inoculation. During that period, and when vaccines were wearing off, animals could still contract and spread FMD but were likely to show only mild clinical symptoms that could easily pass unnoticed. This phenomenon they termed 'masking', and claimed that animals infected in this way could become long-term carriers of the FMD virus and therefore pose an ongoing threat. Their third argument stated that vaccines were far less effective than slaughter. They were useless in young animals, failed to work properly in pigs and sheep, and provided effective immunity against only one of the three types of virus responsible for infection. Finally, they employed national statistics of FMD incidence to demonstrate that nations which adopted a slaughter policy encountered FMD far less frequently than those which vaccinated.

MAF's defenders were quick to point out that neither officials nor FMD researchers were to blame for the deficiencies of FMD vaccines. Rather, in identifying the flawed nature of FMD vaccines and the problems connected with their use, British scientists had shown themselves to be far more advanced than their Continental counterparts, who had yet to notice such drawbacks. They claimed that Pirbright led the world in the quality of its FMD research. It received many visitors from abroad, assisted other nations in diagnosing and controlling FMD, and the director, Dr Ian Galloway, travelled widely on the Continent, forging connections and exchanging ideas with foreign scientists. MAF was well aware of scientists' progress, and was so convinced of the importance of their work that it had decided to improve and extend their research facilities (this version of the story omitted to mention that most of the funding for this project came from the Ministry of Defence, which was interested in furthering its biological warfare research programme; see Chapter 5).[28] Independent research, as advocated by some critics, was both unnecessary and dangerous, as only Pirbright was properly equipped to prevent the highly contagious FMD virus escaping from the laboratory.[29]

These advocates of the slaughter policy alleged that foreign nations envied Britain's relative freedom from FMD and wished that they, too, could stamp out infection using slaughter. However, the prevalence of FMD within their borders made this policy untenable, and so they were forced to adopt the inferior method of vaccination. It was utterly ridiculous for MAF's critics to claim that Britain should follow their example and adopt a 'backward' control method, which would actually worsen the FMD situation. Nor should they listen to Dr Crofton, a well-known quack whose theories had been scientifically disproved by FMD researchers.[30]

There was much truth in many of MAF's arguments. As Chapter 5 revealed, national differences in geography, commerce and scientific tradition meant that that Britain and the Continent experienced FMD in very different ways and chose very different FMD control policies. While the former aimed to eliminate FMD from the nation by means of slaughter and to prevent its re-entry, the latter tried to restrict the spread of endemic disease and to lessen its impact upon meat and milk production. These differences meant that British and Continental researchers judged vaccines in very different ways and reached varying conclusions about their risks and benefits.

In most of Western Europe, where FMD was endemic, slaughter was simply not a policy option: it was expensive, unpopular and provided no guarantee against the reinvasion of disease from neighbouring countries. Of the remaining options for FMD control, vaccines provided far more durable and reliable protection than serum. Farming demand for this technology was extremely high. Consequently, scientists released vaccines for use in the field, prior to exhaustive laboratory testing. They found it impossible, if not unethical, to establish the 'control' populations of unvaccinated livestock, and so were unable to assess accurately the efficiency of FMD vaccination. However, they were not overly concerned by this lack of testing, and believed that the resulting reduction in FMD spread made plain the benefits of vaccination.

In Britain, MAF's insistence upon a slaughter policy meant that scientists could only test vaccines within the laboratory or field station. There they faced a problem rarely encountered abroad – that of replicating 'natural' infection. Placing susceptible animals in contact with the diseased in the hope that they would contract infection produced unreliable results; instead, scientists transmitted FMD by injecting measured amounts of virus into the tongues of test animals. Because all British animals were fully susceptible to FMD, scientists were able to closely monitor the degree and duration of immunity induced by FMD vaccines. These features were regarded as extremely important in Britain, where even a single unchecked case of FMD could lead to a devastating epidemic; but they were ignored in Europe, where the occasional use of ineffective vaccines had little impact upon the overall disease situation. So, while British scientists and officials regarded vaccination as a risky technology that had to be exhaustively tested within the laboratory before it could be pronounced safe, Continental observers had little time for 'artificial' laboratory experiments and presumed that vaccines were safe until proven otherwise.[31]

There were other important reasons why MAF officials and their supporters refused to adopt vaccination. Their private discussions upon the matter show clearly that their policy preferences were underpinned by deep-seated – and perhaps not even consciously recognized – moral, cultural and nationalistic convictions. It seems that they viewed slaughter not only as a method of disease control, but as a moralizing, disciplinary and educational force, and a 'virility symbol' of Britain's superiority over vaccinating nations.

The impact of MAF's FMD control policy was never uniform. Although intended to bring about a 'national benefit' – that of freedom from FMD – the costs of this policy were overwhelmingly borne by individuals living in regions affected by the disease. Unsurprisingly, it was these individuals that proved critical of the slaughter policy (see Chapter 3). MAF branded their complaints 'selfish', and told them that they should quietly bear the hardships associated with FMD control just as in different times and places others had suffered on their behalf. It seems that officials viewed slaughter as a character-building policy that taught self-denial, a 'stiff upper lip' and instilled a sense of the national good. It was also a disciplining policy that required farmers to monitor their animals closely for signs of disease and to adopt certain modes of behaviour to limit disease spread. With two World Wars within living memory, the merits of an obedient, resilient population who put national interests ahead of personal gain could not be overstated, and these aspects of the slaughter policy help to explain why MAF was so reluctant to adopt vaccination. Indeed, in controlling FMD, officials were waging a metaphorical war on disease: FMD was an 'invading enemy' that had to be 'fought' and 'defeated.' Animals had to be 'sacrificed' in the 'campaign' against FMD; MAF veterinary inspectors acted as 'foot soldiers'; the Pirbright institute was the 'British Commando'; and slaughter was 'the first line of defence'.[32]

British proponents of slaughter felt that nations who adopted it were automatically on a higher plane than those who did not because it stamped out germs and gave rise to national disease freedom, a state which MAF had long regarded as the highest possible achievement in livestock disease control. Vaccination, on the other hand, was incapable of defeating FMD because it involved modifying and living alongside the virus. They believed that slaughter could only succeed in disciplined, ordered nations run by enlightened governments. Its effective control of FMD within Britain was testament to MAF's administrative efficiency, Britain's thorough veterinary policing and the educated nature of British farmers – who lived in fear of FMD and were mostly prepared to abide by objectionable disease regulations for the benefit of the nation. But in nations where ignorant farmers did not regard FMD as a dangerous disease and refused to adhere to livestock movement controls, vaccination was the only option.[33] According to this logic, if Britain adopted a policy of vaccination, it would be reduced to the same level as disorganized, inefficient, ignorant and immoral nations such as France, where 'the Latin temperament of the population and the intrusion of politics militate against the effective imposition of restrictions'.[34] Such a move could not, therefore, be contemplated.

THE DISPUTE CONTINUES

MAF's defence of the slaughter policy failed to satisfy its critics, whose complaints continued into the summer and autumn of 1952. Officials blamed public ignorance and 'ill-informed journalism' and increased their efforts, but to no avail.[35] In fact, they mistook the situation. Opponents of the slaughter policy were not ignorant of the issues surrounding FMD control. They disagreed with MAF officials and their supporters because, for them, FMD control was not the purely technical and economic issue that MAF had made out; it was also an ideological affair that was closely bound up with the role and status of science in society, the accountability of government bodies and Britain's international standing.

In effect, the dispute amounted to a clash of worldviews, in which pro-vaccinators fervently criticized those very aspects of the slaughter policy that MAF officials had thought most beneficial. So, while officials adopted a paternalistic view of FMD control, in which a central government body decided what was best for the nation, their critics favoured a more liberal policy that allowed individuals to take responsibility for FMD control and act in accordance with their personal or regional interests. While officials viewed slaughter as an 'advanced' method that demonstrated British superiority, critics thought it an unscientific measure that symbolized Britain's 'backwardness'; and while officials saw vaccination as dangerously uncertain, to critics it provided greater security against the unpredictable invasion and spread of FMD.

Above all, pro-vaccinators thought MAF officials old-fashioned, out of touch and, worse, anti-scientific. The 1950s was a time of unprecedented optimism and faith in the value of science, and as one of the most important achievements of modern medicine, vaccines were celebrated as a symbol of man's mastery over nature and a means of advancing society. To many advocates of scientific enquiry, MAF's criticisms of FMD vaccination just did not make sense. Had it not noticed that in a time of radical agricultural, economic and political change, the traditional policy of slaughter inflicted greater hardships and was less likely to succeed than ever before? How could a 60-year-old policy of slaughter possibly be more scientific than vaccination? How could vaccinating nations be inferior to Britain? And why was MAF so pessimistic about scientists' capacity to develop new and improved vaccines that would one day replace the slaughter policy?[36]

Critics went on to query whether scientists working for a ministry that seemed so unconvinced of the merits of scientific enquiry could achieve anything of value, especially when they seemed more bent on solving foreign nations' problems than their own. The *Daily Telegraph* argued: 'It is not, however, for reasons of national prestige that we have maintained this institute [Pirbright] for the last 30 years and are spending on it this year £93,540. The aim is, or should be, to find a cure or preventative for a scourge.'[37] When, in August 1952, ARC belatedly published a report of research progress. However, this did not have the desired effect of reassuring the public that 'work is being prosecuted with both vigour and competence'. Rather, in discussing vaccines on only 1 out of 12 pages, it confirmed critics' suspicions that scientists had not been concentrating sufficiently upon this important issue.[38] The *Daily Telegraph* complained:

> *The only point of value in it is that the scope of the institute is to be greatly enlarged. . .in that the report is designed to show that the present staff have done their best, it is superfluous; in so far as it implies that this best has been good enough, it is farcical.*[39]

Far from being the 'ignorant laymen' that MAF government officials had assumed, some critics understood the vaccination issue so thoroughly that they were able to detect weaknesses in MAF's pronouncements. They argued that there was no need to maintain complete national immunity against FMD, at an estimated cost of UK£13 million per year, as animals in the vicinity of an outbreak could be ring-vaccinated, once only, against the type of virus causing disease. Alternatively, MAF could permit herd owners to inoculate their animals at their own expense if they wished. They drew on foreign scientists' evidence to claim that vaccines were not dangerous because 'masking' rarely occurred and FMD carriers did not exist. They also argued that MAF's comparison of British and Continental FMD control methods had failed to take into account the indirect financial losses inflicted by slaughter, the psychological benefits of vaccination and the 'teething problems' associated with early vaccine use on the Continent.[40] It seemed, therefore, that MAF's case against vaccination was at best flawed and at worst propaganda. The *Daily Telegraph* complained: 'There seems to be more energy directed towards discrediting vaccination than to trying it.'[41] Farmer George Villiers agreed, stating: 'Never has a case been put before the public so completely from one angle, and I believe, never with so much bias and distortion and obstructive prejudice.'[42] The Honourable R H Bathurst, a cattle breeder, was similarly forthright:

> *In trying to secure a reasonable measure of security for our herds we seem to be fighting something more virulent than virus. We are up against a formidable combination of official ignorance reinforced by personal and professional jealousy.*[43]

These complaints were not without foundation. Even Director Ian Galloway, of Pirbright, felt that MAF had exaggerated the problems associated with vaccination. He wrote a letter to complain that official criticisms had:

> *become more general than is desirable, and have probably arisen from the necessity felt in some quarters so as not to embarrass the maintenance of the slaughter policy, of stressing, rather, the complications and difficulties and impossibilities of producing vaccines of uniformly high protective value.*

He protested:

> *I feel it isn't a good policy, which is adopted in some quarters, to persist in suggesting that figures of disease here compared to the Continent indicate that vaccination is an inferior method compared to slaughter. Too many factors are involved. . . Many of the arguments put up against vaccination are not sound or based on scientific evidence. This has, no doubt, given the impression that any sort of argument is good enough to use against vaccination as a policy, and above all, to prevent any suggestions that the slaughter policy should be abandoned.*[44]

THE GOWERS COMMITTEE OF INQUIRY

In an attempt to dampen down the ongoing controversy, MAF appointed, in August 1952, a departmental committee of inquiry into the epidemic, chaired by Sir Ernest Gowers, an experienced committee man and career civil servant.[45] It directed the committee to focus its attention on the vaccination question, and eagerly anticipated a report that would silence, once and for all, the critics of British FMD research and control. A spectrum of 'stakeholders' were represented on the committee: farming, veterinary, scientific, Labour and the general public.[46] Gowers's preference was for a public inquiry, and Minister of Agriculture Sir Thomas Dugdale agreed, believing that this format would reassure a fractious public. However, the majority within the committee voted for closed hearings.[47] In this they were supported by MAF officials, who felt that as the inquiry addressed questions of policy rather than the exercise of power by a public body, members of the public need not be admitted.[48]

Over the course of 39 meetings, the committee gathered, discussed and critically assessed information drawn from many different and often contradictory sources. It took oral and written evidence from MAF officials, ARC members, veterinary and farming representatives, individuals claiming to possess FMD cures, local authorities and transport associations. Even Dr Crofton was given a hearing, in which he persuaded the committee – against MAF's wishes – to consider further scientific tests on his vaccine.[49] Committee members also visited South America and several European nations, where they gathered information upon alternative disease control policies and learned of foreign experts' views on FMD vaccination.[50]

As MAF had hoped, the committee's report dismissed public criticisms of British FMD scientists and upheld the official belief that there was currently no alternative to slaughter. Nation-wide vaccination was a 'gigantic operation' that was 'manifestly impractical' and would cost UK£24 million, according to a detailed MAF estimate published in the appendix. Vaccines were slow to act, ineffective in certain animals and not completely reliable in others. Consequently:

> *In the circumstances of today, and of the immediate future. . . .any idea that it would be possible to do away with stamping out by making the whole susceptible animal population – or even all cattle – immune by vaccination is in the realms of fantasy. In present circumstances stamping out must continue to be the policy in Great Britain.*[51]

However, all did not go to plan. In the rest of its report, the committee expressed many opinions that deviated from those held by British supporters of the slaughter policy and directly threatened MAF's position. These differences emerged because, unlike MAF officials, the committee assumed that subject to future technical advances, vaccination would eventually replace slaughter as the preferred method of controlling FMD in Britain.[52] It devoted a considerable proportion of its report to discussing available information on vaccines and mapping out vaccination strategies best suited to the British context.[53] In this, it relied heavily upon the evidence of foreign witnesses whose view of vaccination was, at times, blatantly at odds with that held by MAF officials. For example, Chief Veterinary Officer (CVO) Thomas Dalling told the committee that the Dutch held such a low opinion of vaccination that they had recently decided to adopt slaughter.[54] He was later contradicted by Dutch experts, who expressed great faith in vaccines.[55] Dalling also claimed that slaughter had brought the recent Mexican FMD epidemic under control, while Dr Simms, head of the US Bureau of Animal Industry, reported that vaccination had played a valuable role.[56]

Members of the Gowers committee realized, after hearing evidence from many foreign FMD experts, that the view of FMD control which MAF had attempted to impress so dogmatically upon the nation was not universally accepted, and that the costs, benefits and risks of vaccination were by no means well defined.[57] Its report listed many points upon which experts disagreed and, in discussing the lack of evidence on whether vaccines could mask disease spread, went so far as to claim that MAF's fears had been driven by 'a tendency – natural enough in the circumstances – to magnify the unknown'.[58] Although the committee stopped short of defining the exact circumstances under which vaccination should occur in Britain, it reported that during severe epidemics it was prudent to act early, and that ring vaccination in conjunction with slaughter was the most suitable policy. It also recommended that MAF expand the facilities available at Pirbright so that, during an epidemic, it could produce large quantities of vaccine at short notice.[59]

These latter aspects of the Gowers report alarmed MAF officials. In giving credence to the opinions of foreign scientists, they undermined MAF's attempt to present the 'British' view of vaccination as correct, and gave rise to fears that critics – who had long thought MAF's opinions one sided and inaccurate – would declare themselves vindicated. Officials were also horrified by the committee's assumption that MAF would vaccinate in future. Deputy Secretary W Tame wrote:

The committee's recommendation on this point is not quite as forthright as might have been expected. . .the CVO's view is that circumstances would have to be much worse than anything so far experienced this century before he could agree to vaccinate. The committee's recommendations may, however, mean that if we get another epidemic like that of 1951–1952, there will be strong pressure on the ministry from certain quarters to agree to vaccination.[60]

MAF administrator, G R Dunnett, agreed:

I think ministers may wish to be rather stiffer on the question of vaccination. They might, for instance, wish to say that they accept the argument that if the disease got out of control by means of slaughter it would be necessary to consider whether vaccines should be used, perhaps for a short period, to enable control by slaughter to be regained, but that in the government's view we have never approached such conditions and the government see no reason to think that we need anticipate doing so; they must not, therefore, be understood to be accepting any recommendation to include vaccination as part of the methods we use to control the disease.[61]

Tame, Dunnett and their colleagues need not have worried. The committee's report was not published until July 1954, by which time FMD was virtually absent and the earlier epidemic had faded into memory. Having failed to force a change in policy, critics had abandoned their campaign and took little interest in the committee's belated conclusions. MAF was careful not to reignite the debate. Ignoring the more controversial aspects of the report, its press release simply stated that after carrying out extensive enquiries, the independent Gowers committee had dismissed FMD vaccination in Britain as too risky. Slaughter therefore remained the best method of FMD control, although vaccination might be justified during a major epidemic.[62] The report attracted little attention in the press and MAF Parliamentary Secretary Lord St Aldwyn faced few questions when, in July 1955, he told the House of Commons that the government intended to accept its recommendations. Nevertheless, he was careful to emphasize – in accordance with the advice of Tame and Dunnett – that vaccines would only be needed when circumstances were 'exceptionally bad – much worse than anything so far encountered. . .frankly, the contingency seems rather remote'.[63]

MAF's 'cherry-picking' of the Gowers committee's conclusions, together with the public apathy which by now surrounded the issue of FMD control, seriously curtailed the impact of one of the most painstaking, wide-ranging and open-minded enquiries ever undertaken into FMD vaccination. As Chapter 7 shows, for the next 13 years, and even during the devastating 1967–1968 FMD epidemic, MAF officials and their supporters selectively cited the Gowers report in support of existing policy, while ignoring its detailed assessment of the various ways in which vaccination could assist British FMD control. FMD vaccination did not head the political agenda again until 2001 – half a century after it first captured the nation's attention.

Chapter 7

The 1967–1968 Epidemic

SETTING THE SCENE

In the years following the 1951–1952 foot and mouth disease (FMD) epidemic, Britain 'never had it so good'.[1] The austerity of the post-war years passed into history as the economy improved and standards of living rose. Unemployment was low and life expectancy increased. People bought better cars, bigger houses and more sophisticated domestic appliances, and a social revolution began as the younger generation began to assert itself. Overseas, Britannia no longer ruled the waves. It gradually gave up its empire and, humiliated by the Suez crisis of 1956, when international opposition forced the withdrawal of British troops from Egypt, it lost much of its self-confidence. On the European scene, it played second fiddle to France, and suffered the indignity of seeing its 1963 and 1967 applications for European Economic Community (EEC) membership vetoed by French President Charles de Gaulle.

Agriculture also changed as intensive farming took hold. The government continued to believe that increased domestic food production was in the national interest, but aimed for selective growth and greater efficiency rather than outright agricultural expansion. It continued to offer generous farming subsidies, which the Ministry of Agriculture, Fisheries and Food (MAFF) reviewed every year in consultation with the National Farmers' Union (NFU). The 1957 Agriculture Act prevented it from making drastic changes to subsidy levels in the belief that farmers receiving a stable, assured income were more likely to modernize, invest and adopt the latest technologies. Farms grew larger as smaller holdings became unprofitable, and artificial fertilizers, pesticides, purchased feedstuffs, new varieties of crops and animals, the artificial insemination of livestock, and the mechanization of field and farmyard operations all led to enormous growths in output. Most dairy farmers abandoned traditional breeds in favour of the Friesian cow, milk yields increased dramatically and herds grew larger as machine milking took hold.[2]

The same period saw great strides in veterinary medicine. During World War II, the National Veterinary Medical Association (NVMA), led by Harry Steele-Bodger, managed to convince farmers and the government that the veterinary profession had a crucial role to play in preserving the nation's precious meat and milk supply. A MAFF and NFU-backed scheme to provide dairy farmers with subsidized veterinary treatment ran between 1942 and 1950. In guaranteeing many veterinary practitioners a place on the farm, it reversed the fortunes of the profession, which had declined during the inter-war years, as the motor engine replaced the horse. Farmers became persuaded of the veterinary surgeon's capacity to act as 'physician of the farm', and, in the belief that the profession served the national interest, the government provided more funding for veterinary research, backed the creation of a new university-based system of veterinary education, and passed the 1948 Veterinary Surgeons Act, which made it illegal for unqualified individuals to practise veterinary medicine. Meanwhile, the appearance of new drugs such as antibiotics and hormones meant that veterinary practitioners became better at managing disease, and farmers, increasingly mindful of their profit margins, became keener to employ them, not simply on a 'fire-brigade' basis as before, but for the management of herd health and the prevention of disease. At the same time, MAFF's expanded State Veterinary Service made substantial progress in tackling long-standing, costly disease problems, such as tuberculosis (TB) and brucellosis.[3]

But one disease was still at large. Its management had hardly changed in 70 years (see Plate 12) and it was about to cause one of the most devastating epidemics of the 20th century. FMD struck first at a farm near Oswestry, Shropshire, on 25 October 1967. For the next three days there were no more cases, and *The Times* declared the disease 'under control'.[4] But then outbreaks came thick and fast, and one month later, MAFF veterinarians were diagnosing up to 80 new cases each day. FMD incidence dropped during December and January; but the disease was not completely stamped out until June 1968. Altogether, there were 2228 outbreaks, 94 per cent of them in the north-west Midlands and north Wales. Nearly half – 1021 outbreaks – occurred in Cheshire and 727 in Shropshire. The slaughter tally and the cost of disease elimination were unprecedented. 450,000 livestock lost their lives, among them one third of the cows in Cheshire. Compensation cost UK£27 million, with total losses estimated at between UK£70 million and £150 million.[5]

The impact of this epidemic was felt at many levels. As in past epidemics, the methods used to control FMD affected the work and lives of thousands of rural inhabitants. The disease also made its presence felt within urban areas, where for the first time a broad section of the populace saw their leisure pursuits curtailed. For MAFF officials, the rampant spread of infection caused a crisis of confidence in the traditional slaughter policy, and, in agreeing to impose a temporary ban on meat imports from FMD-infected nations, the government precipitated a breakdown in Anglo–Argentine relations, which had ramifications for Britain's export industries. But while the effects of FMD were experienced in many different ways by a great variety of people, on one matter there was unanimity: all who had been touched by the disease vowed that such an epidemic must never happen again.

But as everyone knows, FMD reappeared in 2001 and wreaked greater havoc than ever before. The history of 1967–1968 provides an important backdrop for that event (see Chapter 8) and reveals just how few of its lessons were remembered 33 years later. It also draws together several issues raised earlier in the book, such as the intense suffering inflicted by the official FMD control policy, the highly political nature of disease control, the disparity between MAFF's public pronouncements and its private actions, the difficulties involved in controlling the importation of FMD-infected meat, and the institutional resistance to vaccination. Once again, FMD was far more than just an animal disease.

THE VIEW FROM THE GROUND

When FMD broke out near Oswestry on 25 October, MAFF's veterinary department was relatively unconcerned. It had already overseen the elimination of 33 outbreaks that year and had no reason to believe that this would be any different.[6] Regional veterinary staff went through the familiar routine of supervising the destruction and burial of infected stock and tracing the animals with which they had come into contact, a somewhat difficult job since two cows had been sent from the infected farm to Oswestry market, along with another 7000 animals. Nevertheless, by 30 October, the task was complete. No further cases of FMD had come to light, and so MAFF lifted the restrictions that it had automatically imposed upon the movement of livestock in the surrounding counties. But, then, two new notifications were received from Cheshire and North Lancashire. Others followed. By 1 November, 19 outbreaks of FMD had been identified and over 4000 animals destroyed. Six days later, the number of outbreaks reached 106, most of them within the north-west Midlands.

As slaughter and burial progressed, the Oswestry area became 'a landscape without life. Mile after mile of fields, which should be full of sheep and cattle, are empty.'[7] For those farmers who had lost their stock, it seemed that little had changed since their grandparents' day. As one commentator noted: 'They use an automatic pistol to slaughter the cattle instead of a pole-axe, and a mechanical digger for the grave instead of men with spades. . .not a very good advertisement for 40 years of ministerial research.'[8] For the second time in half a century, rural life in Cheshire ground to a halt. Farmers barricaded themselves inside their properties, laid disinfectant-soaked straw on roads outside, ordered grocery deliveries to be dropped at the farm gate, and hung up notices warning people to 'Keep out!' 'Stay away!' and 'Be off or be shot!' Towns became ghostlike as markets and social gatherings were cancelled, pubs emptied and children stopped going to school.[9]

Where possible, farmers tried to follow MAFF and NFU advice about housing animals, moving them away from farm boundaries, disinfecting milk churns before and after collection, and avoiding contact with other livestock owners. But in spite of everything, the disease continued to spread, attacking rich and poor, pedigree and commercial herds alike. Every week, the *Chester Chronicle*'s list of infected farms grew longer. Successive press reports described the epidemic as 'the worst since 1966', 'the worst since 1960. . .1952. . .1923. . .this century. . .since the slaughter policy began. . .ever recorded'. Then it was likened to the 1865 cattle plague epidemic; finally, it became 'the plague to end all plagues', 'the biggest loss and crisis ever to hit the livestock industry of Great Britain'.[10]

A chain of regional control centres sprang up as FMD spread to new areas of the country. Each was headed by a MAFF regional veterinary officer (RVO), with the assistance of a lay regional controller. Veterinary surgeons took charge of disease control and delegated the day-to-day running of the centre to lay staff who kept in touch with the local authorities, police, and NFU branches. It soon became clear that the centres were grossly understaffed, so MAFF drafted in veterinarians from nearby practices, members of the Royal Army Veterinary Corps, international volunteers and staff based at other MAFF centres and laboratories. Over 500 were temporarily enrolled in the State Veterinary Service between 25 October and 8 December. Based far from their homes and families, they worked long days, carrying out extremely unpleasant tasks under intense pressure.[11] The *Chester Chronicle* had nothing but praise for their hard work and dedication: 'They are essentially humane men, and for them to have to turn. . .to wholesale destruction cannot but be a revolting exercise to which they must steel themselves.'[12]

Various logistical problems hindered the control of FMD, and after just 12 days, MAFF had no alternative but to ask the army for help.[13] Slaughter men were in short supply, there were disputes over pay and, reportedly, many labourers left premises without disinfecting themselves. Maps and forms were unsatisfactory, new staff needed training and local accommodation was often poor. Burial, the preferred method of carcass disposal, required permission from the water authorities, and many carcasses had to be burned, a more unpleasant and costly method. There were shortages of heavy machinery, clothing and pressure sprayers for disinfection; and overloaded phone lines led to communications difficulties. Several of these problems would not have arisen had regional control centres made full use of local authority resources, and their failure to do so drew much criticism.[14] MAFF later looked into its arrangements for managing the disease and acknowledged that it had not been well prepared. Nevertheless, officials argued that 'On the basis of experience in the years before October 1967, it would have been unreasonable for the ministry to have built up arrangements for coping.'[15]

An additional pressure, which took MAFF by surprise, was the public and media interest in the disease. As one official noted: 'From the outset the demand for information was overwhelming.'[16] At Oswestry, officials joined with the NFU to establish a separate information centre that issued bulletins to local radio and TV stations and held a daily press conference at which a veterinary officer answered questions. Staff numbers were increased as phone lines grew busier, and reporters and cameramen were in constant attendance until mid December. MAFF's veterinary headquarters at Tolworth also set up a press and general enquiry point, which issued bulletins direct to the BBC and Press Association, and released press notices detailing changes to the infected and controlled areas. Such information helped to keep people abreast of the changing disease situation, and impressed upon them those modes of behaviour required to minimize the risk of FMD spread. Most media coverage was uncritical in tone, and officials later praised journalists' 'responsible reporting' of the epidemic.[17] The situation was very different in 2001 (see Chapter 8).

On 18 November, by which time 495 cases of FMD had been confirmed and 93,000 animals slaughtered, the whole of England and Wales became a 'controlled area'. Markets for store stock were forbidden and the holding of fat-stock markets required a local authority licence, usually issued by the police. One week later, Scotland became subject to the same restrictions. Within infected areas (anywhere within 10 miles of a confirmed outbreak), all animal movements had to be licensed. Because autumn was one of the busiest times of year for shifting livestock around, these restrictions caused widespread disruption and the price of meat soon rose.

Farmers and veterinary surgeons alike pondered the 'unprecedented' speed of FMD spread and its curious 'jumping' between widely separated farms. Some blamed starlings for carrying the virus on their feet; but

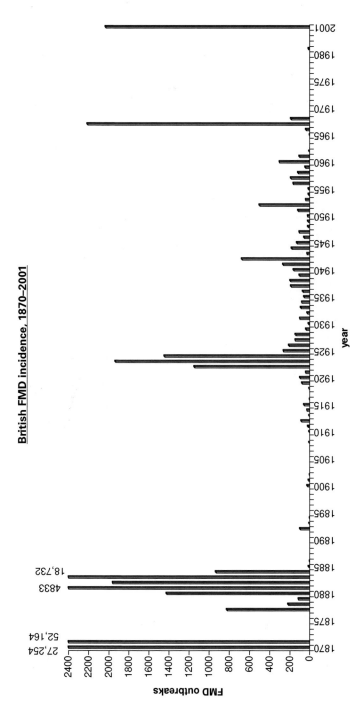

Plate 1 *Graph of foot and mouth disease incidence, 1839–2001*

CATTLE DISEASES PREVENTION BILL.

63, WEST SMITHFIELD,

March 5th, 1864.

I am directed, as Chairman of a Committee appointed at a Meeting of Graziers, Salesmen, and others interested in the Cattle Trade, to call your attention to a Bill now before Parliament, entitled, "*A Bill to make further Provisions for the prevention of Infectious Diseases amongst Cattle,*" which stands for a second reading on Wednesday, the 9th inst., and which seriously threatens the interest of all persons engaged in the rearing and feeding of Cattle, as well as in the sale thereof.

The Committee views with much alarm the stringent Clauses of this Bill, particularly as referring to the disease known as the " Foot and Mouth Complaint," which, in their opinion, if carried out, will affect the supply of food to the people to an extent not contemplated by the Legislature, and will operate oppressively and even ruinously upon individuals connected with the supply of such food.

The Committee also begs to draw your attention to the fact that the existence of the " Mouth and Foot Complaint" for so long a period as twenty-five years, without any perceptible or injurious influence upon the health of the people of this country, leads to the inference that this complaint has not the effect of rendering the flesh of animals unwholesome ; which opinion is supported by experience, as well as by the highest professional authorities.

The Committee further desires to impress upon you that the character of the " Foot and Mouth Complaint " shows that it should not be legislated for as a purely contagious disease, it being an established fact, that animals which leave the premises of their respective owners in a state of perfect health, are often affected by this epidemic before they reach their destination, and this solely from an atmospherical influence.

Trusting, therefore, that you will use your influence so to modify this Bill, either by obtaining the withdrawal of the disorder in question from the Schedule of diseases, or by having the Bill referred to a Committee empowered to call evidence,

I have the honour to subscribe myself,

Your most obedient and humble Servant,

JOHN GIBLETT.

To

Plate 2 *John Giblett, leaflet on the 1864 Cattle Diseases Bill*

Source: Reproduced with permission of the Archives and Historical Collection of The Royal Veterinary College, London

Plate 3 *J B Simonds*

Source: Reproduced with permission of the Archives and Historical Collection of The Royal Veterinary College, London

Plate 4 *George Brown*

Source: Reproduced with permission of the Archives and Historical Collection of The Royal Veterinary College, London

COUNTY OF LINLITHGOW

FOOT AND MOUTH
DISEASE

THE LOCAL AUTHORITY of the COUNTY OF LINLITHGOW, in virtue of the powers conferred upon them by "The Contagious Diseases (Animals) Act, 1878," and "The Scotland Movement into District (Foot and Mouth Disease) Order of 1882," with the view of preventing the introduction of Foot and Mouth Disease into the County, do hereby make the following Regulations, viz:—

I. No animals shall be moved into the District of the Local Authority of the County of Linlithgow without a License from the Local Authority of the County, which License shall not be granted until satisfactory evidence has been adduced to the Local Authority of the County that the animals to be moved—

(1.) Either have been bred on the farm or premises from which they are to be moved or have been pastured or fed thereon for a period of at least twenty-eight days before the granting of the License;

(2.) Have not, within twenty-eight days immediately before the granting of the License, been exposed in any market, fair, exhibition, or public sale; and

(3.) Are not affected with Foot and Mouth Disease, and have not been affected with that disease nor been in contact with animals affected with, or suspected to have been affected with, that disease, within the said period of twenty-eight days.

II. Licenses shall be available only for Four Days after the same shall have been granted, and shall, within Two Days after the animals therein mentioned have been moved into the County, be delivered by the person to whom it is granted to a Constable of the Linlithgowshire Police.

III. The Clerk to the Local Authority of the County is hereby authorised to issue Forms of Declaration and grant the necessary License upon production of the evidence above specified.

IV. These Regulations shall take effect from and after the 17th day of February 1883, and shall continue in force until revoked or altered.

BY ORDER,

ROBERT R. GLEN, Clerk to the Local Authority.

LINLITHGOW, 16th February 1883.

NOTE 1.—The above Regulations do not restrict movement of Animals by railway through the County of Linlithgow.

NOTE 2.—If an Animal be moved in contravention of the above Regulations, the owner thereof, and the person causing, directing, or permitting the removals, and the person or company moving or conveying the Animal will be liable to prosecution.

NOTE 3.—It is an offence against the Act of 1878 to send or carry, or cause to be sent or carried, on a railway, canal, river, or inland navigation, or in a coasting vessel, or on a highway, or thoroughfare, any fodder that has been in a place infected with Pleuro-Pneumonia, Foot and Mouth Disease, Sheep Pox, or Swine Fever, or that has been in any place in contact with, or used about a diseased animal, horse, ass, or mule, except with a license of the Local Authority for the district in which such place is situate, on a certificate of an Inspector certifying that the fodder moved has been as far as practicable disinfected.

Plate 5 *County of Linlithgow, Foot and Mouth Disease Bill, 1883*

Plate 6 *Group of Royal Irish Constabulary at Swords, 1912*

Source: Department of Agriculture and Technical Instruction for Ireland, *Report on FMD in Ireland in the Year 1912*, pp1914, cd 7103, xii, 793

Plate 7 *'Mouthing' a suspected animal, Ireland, 1912*

Source: Department of Agriculture and Technical Instruction for Ireland, *Report on FMD in Ireland in the Year 1912*, pp1914, cd 7103, xii, 793

Plate 8 *Disinfecting before leaving an infected place, Ireland, 1912*

Source: Department of Agriculture and Technical Instruction for Ireland, *Report on FMD in Ireland in the Year 1912*, pp1914, cd 7103, xii, 793

Plate 9 *Disinfecting before leaving, 2001*

Source: Silence at Ramscliffe. Copyright © Chris Chapman 2001

Plate 10 *Sir Stewart Stockman*

Source: Journal of Comparative Pathology and Therapeutics, 1926

Plate 11 *Policeman at the gate, 1938*

Source: Reproduced with permission of Rural History Centre, University of Reading

Plate 12 *The scene at Sturminster Newton, 1935*

Source: Reproduced with permission of Rural History Centre, University of Reading

Plate 13 *Research facility, Pirbright, circa 1924*

Source: Courtesy of Institute for Animal Health, Pirbright

Plate 14 *Funeral pyre, Lower Dalby, Yorkshire, 1958*

Source: Reproduced with permission of Rural History Centre, University of Reading

Plate 15 *FMD disruption, 1967*

Source: Reproduced with permission of the *Farmers' Guardian*

Plate 16 *'Swill must be boiled' MAFF education poster, 1960s*

Source: Reproduced with permission of the National Archives

ornithologists rejected the idea. Others suggested that the wind had carried virus across the Cheshire plain, a theory confirmed later following the analysis of epidemiological and meteorological data. But to most observers, it was clear that intensive farming practices were at least partly to blame. Farms were larger and more densely stocked than in the past, so each outbreak affected more animals, generated a larger volume of virus and resulted in a higher slaughter tally. The increasing scale and frequency of livestock movements further contributed to FMD spread, as did modern practices such as the collection and transport of milk in bulk tankers, the spraying of slurry and artificial insemination.[18]

Within the infected areas, MAFF used its statutory powers to close footpaths and halt country sports, such as angling, shooting, hunting and racing (see Plate 15). Minister of Agriculture Fred Peart also appealed to the public to stay away from the countryside and not to participate in climbing, walking, boating, coursing, motor racing and rallies, as well as outward bound courses. For events involving large gatherings of people in rural areas, officials sought veterinary advice upon the disease risk involved and, where necessary, recommended the relevant organizations to halt or curtail their activities. Through such restrictions, FMD impacted upon the lives of individuals far removed from agricultural circles.[19] For some, however, Peart had not done enough. The NFU and the Tory opposition wanted a more 'cut-and-dried' set of guidelines and criticized Peart's refusal to follow the example of Irish President Blaney, who had halted all rural leisure activities throughout Ireland for fear of FMD spread from the mainland.[20] The mounting pressure compelled MAFF officials to reconsider their policy in the light of political rather than veterinary concerns. They subsequently banned all hunting and horse racing, and introduced a new FMD (Temporary Restrictions) Order, which granted Peart the power to halt events in agricultural areas. They also considered whether Peart should seek emergency powers to forcibly restrict access to the countryside, but decided (in stark contrast to 2001) that such measures would cause unjustified disruption.[21]

Early in December, Peart announced that he would set up an independent inquiry into the epidemic and an internal investigation into its source.[22] By then, the number of new FMD cases notified each day had begun to fall, and soon it became clear that the epidemic was in decline. There was little celebration in Cheshire, however, as thousands awoke each day to silence and an empty farmyard. The numbers afflicted increased, day by day, as more farms were slaughtered out. Those who had so far escaped the infection remained in their homes to wait and pray. As in 1923, local newspapers began to voice fears that the county would never recover from the blow struck by FMD.[23] The editor of the *Chester Chronicle* wrote:

> *Along the high roads, braziers glow at farm gates and deep in the country, the ruthless slaughter goes on, signalled by fire and stench and drifting smoke. . . It is a sad, grim, desperate business. No other industry, or way of life, suffers like this, and it takes a great deal of nerve to stand it.*

He also told of the 'Farmers and their wives and workers [who] write to us in a way I have never before known. They tell us on the phone of their losses and fears and shattered futures. They seek not sympathy so much as a shared understanding of their misfortune.'[24]

Several farmers told their stories to the national press. One man, who had lost his stock twice before in 1952 and 1961, wrote:

> *I have no hope. . . You think the bottom has dropped out of your world. You feel like doing nothing at first. The first night I did not sleep at all, just thinking about what was in front of me – for the third time. You wake up in the morning and prepare yourself, but there is nothing. Every day is black. Every neighbour has had his herd slaughtered.*[25]

Another related his sorrow and anger at the city-dwellers' failure to comprehend the rural plight:

> *You listen on the wireless: 'Numbers down, only 45 – all in the controlled area.' Only 45. No one who isn't tied and connected with the land can understand the enormity of the disaster which is now with us – you walk, you drive through a countryside which you do not realize is locked in battle, which is under siege. . . You don't love a herd – you can't love 150 animals – but they are a symbol of your achievement; they are your fulfilment. And when that's snuffed out in this brutal fashion, well, you die a little. . . The cost will be calculated one day – in cash, but never in human tragedy. . . At least if you got the disease, waiting would be over; but a pressure builds up and is with you day and night. . . And, oh, how I wish the general public could under-stand how we, the Cheshire farmers, feel when we hear that bold and terrible phrase, 'Only 45 today.'*[26]

Under normal circumstances, farmers could restock their farms four to six weeks after slaughter ended. But because the entire country was still a controlled area, it was impossible to move livestock around. Some farmers were insured against consequential losses; but the remainder had to survive long periods without income, and many could no longer afford to pay their workers. The scale of the slaughter aroused fears that when the nation was released from restrictions, the demand for new stock would push up prices and encourage the sale of TB or brucella-infected animals. Indeed, to the disadvantage of farmers who had lost their animals at the start of the epidemic, compensation pay-outs had already crept up. MAFF tried to address these issues. Working alongside the NFU and breed societies, it drew up lists of farmers who had stock for sale. It paid farmers to disinfect their property after the slaughter and found places for many agricultural labourers within its slaughter and burial teams. On condition that the industry contributed half of the cost, it offered top-up compensation to farmers who had lost their stock during the first four weeks of the epidemic. It also announced a 'ploughing-up grant' for farmers who decided to delay restocking and convert their land to arable. Other benefits included a temporary tax exemption on compensation pay-outs and no loss of subsidies for slaughtered animals.[27]

By the end of December, new outbreaks had dropped from a peak of over 80 to between 10 and 20 a day. The situation continued to improve throughout January and February, causing MAFF officials to redraw the boundaries of infected and controlled areas and reassess the trade and leisure restrictions required within them. This proved an extremely difficult task. Farmers fearing the spread of FMD wanted measures to remain in force; but rural inhabitants whose businesses had suffered pressed for their removal.[28] On 4 January, E H Bott, MAFF under-secretary for Animal Health Division I (AHDI), discussed the issue in a memo, amusingly entitled 'Huntin' Shootin' n' Fishin'':

> *From a veterinary viewpoint it seems to us here in AHDI that there might be risks involved if persons living in contact with livestock in infected areas contacted livestock in other areas: and no more than this is involved. The rest is a piece of window dressing. But let us try to dress the window sensibly.*[29]

Bott recommended the release of large swathes of the country from restrictions; but MAFF Deputy Secretary W Tame released only selected areas, saying: 'We would never be forgiven if we relax too soon.' Bott complained that his plan had been based upon expert veterinary advice and was: 'What I myself believe is needed on political grounds, if we realize that the NFU constitute only a small minority of the population of this urban country.' Another MAFF under-secretary, J Carnochan, later protested: 'Any benefits from maintaining controls longer than needs be are necessarily speculative; the disadvantages do not have to be speculated on for they are plain to see.' However, Tame refused to lift the controlled area restrictions until the very end of February.[30]

Slowly, farmers began to restock, and by the end of March over 1300 had done so, aided in Cheshire by the Foot and Mouth Restocking Association, which negotiated prices on their behalf. Alarmingly, a handful of farms succumbed to the disease a second time, and, blaming the presence of infected hay, MAFF veterinarians made arrangements for all hay and loose feed on empty premises to be sprayed with formalin.[31] But, gradually, life in the rural midlands returned to normal, and on 13 March, Oswestry cattle market opened for the first time in four and a half months. The following day saw the first meeting of the committee of inquiry into the epidemic, chaired by the Duke of Northumberland.[32] But the epidemic was not yet over. Occasional outbreaks continued, and MAFF did not declare the nation free of FMD until 25 June 1968.

THE VACCINATION QUESTION

The 1967–1968 epidemic again raised the question of FMD control by vaccination. Proponents of this policy believed that it was a more moral, scientific and economic method than slaughter and considerably better suited to modern farming conditions. However, such views – though seemingly prevalent at grassroots level – received little publicity outside the correspondence pages of newspapers.[33] Almost all journalists and politicians expressed a preference for slaughter, even when it became apparent that FMD was running rapidly out of control. Consequently, there was no re-run of the earlier controversy described in Chapter 6.

There are several explanations for this turnaround in public opinion. It seems that people no longer expected – as they had in 1952 – that vaccine discovery would lead to the replacement of the slaughter policy. In almost every sizeable epidemic since then, MAFF and its supporters had advertised the many problems associated with vaccination. They had cited the Gowers committee estimate that vaccination would cost UK£15 million a year, highlighted its recommendation that slaughter should continue and emphasized the dangers of vaccination and its many scientific deficiencies. In fact, since the 1951–1952 epidemic, scientific and technological progress had considerably reduced the risks and cost of FMD vaccination; but anti-vaccinators took little notice and, by force of repetition, their arguments had the desired effect upon public opinion.

The case for vaccination was weakened further by the successes of the traditional policy, which left the nation free of infection for almost three years during the early 1960s. Also, many continued to support slaughter for nationalistic reasons, seeing it as a sign of Britain's superiority over vaccinating nations. Britain could 'boast' of its FMD-freedom and had 'gone to great lengths and expense to remain the healthiest livestock country in the world'. MAFF's professional staff and the research centre at Pirbright were 'the best in the world'. To vaccinate would be 'a national calamity', for it was still a 'last resort', a 'second line of defence' or a 'fall-back'. It would 'ruin Britain's role as a producer of disease-free livestock', and was 'tantamount to resigning our proud freedom from disease'. As one Cheshire veterinary officer pointed out: 'We are fighting for the health of British agriculture and when we win through the slaughter of so many of our stock [it] will not have been in vain.'[34]

As far as the general public were concerned, vaccination was not the dominant issue of the 1967–1968 epidemic. Although much discussed in the press and in Parliament, there was little public or political pressure for its adoption. For MAFF's veterinary department, however, vaccination assumed unprecedented importance as November drew on and the disease situation worsened. In public, Peart and his veterinary staff continued to reassure the nation of the forthcoming victory over FMD; but, in private, there was a growing sense of anxiety and even panic as they realized that the unimaginable might happen and the slaughter policy fail. Eventually, MAFF's senior veterinary staff decided that they must draw up plans to vaccinate in the north-west Midlands.[35] This was a momentous concession. Never before, except under threat of biological attack, had MAFF conceded a role to FMD vaccines in Britain. When he heard the news, Dr Noel Mowat, a scientist based at Pirbright, told a colleague: 'We are living in history; the ministry are going to vaccinate.'[36]

Despite the significance of this move, it received little publicity. Whether for fear of angering the NFU, arousing the hopes of anti-slaughter campaigners, or perhaps because MAFF was still extremely reluctant to admit that its age-old policy might fail, Peart deliberately played down the plan to vaccinate. He informed Parliament not in his statement of 27 November, which discussed the progress of the epidemic and received widespread publicity, but the following day in a concise reply to a parliamentary question. He emphasized that vaccination would provide 'a second line of defence' against the disease, and described his decision to acquire vaccine from overseas as 'purely a precaution. I am still convinced on the basis of the advice of my professional staff – the best in the world – that the slaughter policy is in the best interests of the country.'[37] Later, in a BBC broadcast, he repeated the usual economic and scientific arguments against vaccination, and said that he would use it only 'as a last resort. . .when the present control campaign completely breaks down'.[38] This approach had the desired effect; there was no surge of public interest and the news commanded little attention in the press.

In the days that followed, Peart continued to reaffirm his faith in the slaughter policy and to emphasize the problems associated with vaccination. Behind the scenes, however, preparations to vaccinate were well underway. Three million doses were flown in from France, under tight security in case they fell into the wrong hands, and another 2 million doses were purchased from the Wellcome laboratories in South America. [39] On 28 November, the CVO, John Reid, summoned Mary Brancker, president of the British Veterinary Association (BVA, formerly the NVMA), to a meeting and together they planned the vaccination campaign. [40] She later explained:

> *We calculated the number of animals to be vaccinated, the number that could be examined and vaccinated in an hour and the number of hours of daylight available in December. These calculations gave us the number of veterinary surgeons required . . .finally, we agreed the rate of pay for them.* [41]

On 2 December, as MAFF press officers scotched rumours that vaccination was about to begin, senior staff at the Oswestry control centre met to devise a blueprint area vaccination scheme. Their recommendations formed the basis for a memo, sent from MAFF's veterinary headquarters at Tolworth to senior veterinary and lay staff in the field. It detailed how vaccination would be carried out in the Cheshire and Shropshire areas, and covered staffing arrangements, paper work and vaccine supplies. It emphasized: 'this. . .is not to be interpreted as an indication that any such decision [to vaccinate] has been or is likely to be taken. The work is purely precautionary; but it should be put in hand at once.' [42]

By the time the memo was issued, on 7 December, FMD incidence had begun to fall. Shortly afterwards, senior MAFF veterinarians at Tolworth judged that slaughter would win through after all, and they abandoned plans to vaccinate. Once again, MAFF had managed – this time by the skin of its teeth – to maintain its record of never having vaccinated against FMD. But official confidence in the slaughter policy had been shaken. As George Amos, MAFF regional controller for the north-west region, noted: 'I wonder whether the disease might have got away completely. Was it a damned near thing? What are the risks nowadays of an epidemic of a comparable or even larger scale happening again?' [43]

Realizing that, in the short term, public opinion would not tolerate a rerun of the 1967–1968 epidemic, officials decided to keep the imported vaccines on hand in case FMD reappeared; but, thankfully, it did not. In the long term, however, they were less convinced of the need for vaccination. Most continued to view it as a costly, scientifically unsound policy, and Tame even believed it psychologically damaging to farmers to have 'compulsory vaccination imposed on them with liability for prosecution and fines if they do not comply'.[44] They submitted evidence along these lines to the Northumberland committee and put future policy considerations on hold until it reported.[45]

TO IMPORT OR NOT TO IMPORT? THE MEAT QUESTION

The 1967–1968 epidemic was only days old when commentators began to ask the perennial question: where had FMD come from? As the disease situation worsened, this issue increasingly dominated the political agenda and brought to a climax the 40-year-old controversy over British meat import policy. As Chapter 4 showed, farming agitation for a ban upon meat imports from FMD-infected countries began with the 1926 discovery that meat could carry the FMD virus. The British government resisted such calls on account of the consumer demand for cheap meat, but tried to encourage meat-exporting South American nations, most notably Argentina, to take more effective action against FMD. However, because the two nations experienced FMD in very different ways, and had different political goals, commercial interests and cultural perspectives, they found it extremely difficult to agree upon the correct manner of FMD control.

British farmers' demands for an import ban died away with the outbreak of World War II, only to re-emerge during the later 1950s when meat rationing ended and FMD incidence rose.[46] In the intervening years, British meat production and imports from FMD-free nations had risen, while under populist President Juan Peron, Argentine meat exports had fallen to approximately half of their pre-war levels.[47] This new state of affairs made an import ban more feasible than ever before. Meanwhile, news from Argentina gave strength to farmers' claims that its meat posed an unacceptable disease risk. The Bledisloe agreement of 1928 (which was designed to prevent the export of diseased meat to Britain; see Chapter 4) was not working properly because there were insufficient Argentine veterinary surgeons to examine livestock on ranches and at slaughterhouses for signs of FMD. Vaccination was underway, but had had little impact on disease incidence because vaccines were not used in a systematic manner, supplies were limited and quality control was non-existent.[48]

During the late 1950s, the CVO, John Ritchie, tried again to persuade the Argentines to improve their FMD controls, as did Anthony Hurd, chairman of the Conservative Agricultural Committee. They met with little success, mainly because the Argentine government refused to acknowledge that FMD was not under control. Hurd blamed 'Argentine pride' and the 'Argentine temperament' (which he likened to that of the French!), but commercial and political factors were also involved.[49] Argentina was still heavily dependent upon meat exports to Britain, which provided 40 per cent of its sterling earnings, and the threat that FMD posed to the national economy was well recognized. Nevertheless, cattle producers and *frigorifico* owners resisted the introduction of additional disease controls, partly on grounds of cost, and partly because many still disagreed with the British belief that FMD was a serious plague, and rejected the 'unproven' link between their meat and British FMD outbreaks. The succession of weak governments that followed in Peron's wake were unable to overcome this resistance and could do little to preserve the threatened trade other then deny the existence of FMD. [50]

Although, in public, MAFF spokesmen maintained their traditional resistance to farmers' demands, in private they tried to drum up the necessary government support for restrictions upon the meat import trade.[51] However, they made little headway until 1960, when Britain suffered the worst FMD epidemic since 1951–1952. Imported meat, bones and swill were strongly implicated (see Plate 13), and MAFF gained permission to ban pork imports from nations where FMD was endemic.[52] In Argentina, pork did not carry the same cultural significance as beef. Nevertheless, the government was extremely unhappy with the ban, and decided, at long last, to start a more effective FMD control programme. It empowered the minister of agriculture to enforce the compulsory vaccination of cattle, appoint additional veterinarians and establish zones within which vaccination of all species was required. Slowly, the disease situation improved. Concurrently, Britain experienced its longest period of FMD freedom for over 60 years (June 1962 to April 1965), and farming calls for trade restrictions died away.[53] But this was merely the calm before the storm. By the autumn of 1967, Britain was once more in the grip of FMD, and it was not long before blame descended upon imported meat.

As usual, farmers led calls for a ban upon meat imports from FMD-infected countries, and they became increasingly vocal as the disease situation worsened. Butchers and meat traders naturally opposed their demands and lobbied for the maintenance of the trade.[54] The issue arose frequently in Parliament, where Peart denied all knowledge of the epidemic's cause and refused to commit himself to action against the meat import trade.[55] Once again, though, MAFF's activities behind the scenes told a very different story. Although the evidence was far from concrete, officials had, from the start, suspected the involvement of South American meat, and on 23 November, Peart suggested to fellow ministers that he halt the trade. They were none too enthusiastic, and so he agreed to seek more definite evidence of disease origin and report again in a week. In the meantime, the Argentine foreign minister summoned the British ambassador to Buenos Aires, Sir Michael Cresswell, and warned him of the 'grave consequences' that would follow any interference with the meat trade.[56] On hearing of this thinly veiled threat, the Foreign Office became extremely alarmed. Anticipating that the Argentines would probably retaliate against the British export trade and stymie concurrent negotiations over the future of the Falkland Islands (which the Labour government wanted to hand over to Argentina),[57] it planned to resist Peart's call for a trade ban.[58]

However, by the next Cabinet meeting, Peart had changed his rationale. He no longer wanted to ban South American meat imports because they had caused the current crisis; he simply wanted to ensure that FMD did not reappear at a time when MAFF's veterinary resources were operating at full stretch. This was a more persuasive argument. With over 70 new cases of disease reported every day, the last thing Cabinet members wanted was another epidemic. For the first time in the 40-year history of controversy over infected meat imports, MAFF's request for a trade ban gained the necessary government support. However, at the insistence of the Foreign Office, it was agreed that the embargo should last only until the epidemic came under control or for three months at the most. Furthermore, to temper the inevitable Argentine outcry, it should apply to all FMD-infected nations. On 4 December 1967, the House of Commons learned of the Cabinet's decision. Farmers were extremely pleased, although there followed 'great pressure by correspondence and telephone from the meat trade'.[59] Inevitably, Peart was asked why he had waited until the epidemic was a month old before taking action. The *Daily Telegraph* commented: 'The case. . .is far from made out, and the suspicion must remain that Mr Peart has yielded a doubtful point under pressure.'[60]

Peart was careful to emphasize that the trade embargo did not imply that imported meat had caused the epidemic; but few believed him, especially in Argentina. The Foreign Office's fears were quickly realized as the day after the trade embargo was imposed, Argentine buyers announced their intention to boycott Newmarket horse sales, where they were expected to spend UK£250,000.[61] Over the next few weeks, news gradually filtered through of unexplained delays in the signing of public contracts that the Argentine government had provisionally awarded to British firms dealing in road-building equipment, aircraft, submarines and telecommunications equipment. When tackled on the issue, Argentine government representatives denied all knowledge of retaliatory practices. However, the British embassy in Buenos Aires later discovered that an instruction to stop signing or awarding contracts to British firms had come from the very top, from President Ongania's office.[62]

These measures hit home as a result of the concurrent crisis in the British economy. In November 1967, the government recorded the worst ever peacetime trade deficit and announced the devaluation of the pound. Solving Britain's balance-of-payment problems hinged upon expanding exports, and Argentina, where the economy was recovering after years of crisis, had been viewed as a promising market.[63] Such hopes were now dashed. Argentine officials took further advantage of the situation by suggesting that the two nations open negotiations upon the potential expansion of Anglo–Argentine trade, and later indicated their willingness to spend UK£180 million on British goods, an offer that was deliberately leaked to the British press.[64]

The Argentine response to the meat import ban is understandable when one considers the commercial and cultural importance of its meat trade, and it should have come as no surprise to British government officials who had learned, from the events described in Chapter 4, of its tendency to react violently against proposed and actual trade restrictions. As already described, the two nations were bound by a complex set of commercial and historic ties. In an attempt to preserve these links, the British government had, during the mid 20th century, exempted South American nations from the rising standards of animal and public health that it had gradually imposed upon other meat-exporting countries.[65] Having acquired this privileged position, the Argentine government did not intend to relinquish it without a fight.[66] Its retaliatory response to the meat embargo transformed FMD from an agricultural into a diplomatic and a commercial issue, and forced the intervention of British government departments other than MAFF. Such tactics were relatively successful, as we shall see.

As the new year dawned, MAFF officials met to discuss the future of the trade ban. Most wanted it to remain in place on animal health grounds. Deep down, however, all realized that the conditions under which it had been imposed and the reactions that it had provoked made retention politically impossible. They were also aware that on resuming trade they would be criticized for exposing the nation to disease. Until the Northumberland committee of inquiry reported, they could not finalize future meat import policy; but this would not occur for some months as the committee's membership had not yet been decided.[67] So, how, in the meantime, could they reconcile Argentina's renowned sensitivity towards its meat with the demands of British farmers, consumers, meat traders, foreign policy-makers and the exporters of manufacturing goods?

On consideration, MAFF officials decided that their best option was to send a high-level British veterinary mission to South America to investigate the workings of its FMD control legislation and to recommend future safeguards upon the meat import trade. All acknowledged that the true purpose of the mission was 'window dressing' (not the first time this phrase had been used!). It could not possibly overcome the main obstacle to Argentine FMD control, which was its small and ineffectual veterinary service. It could, however, reduce criticisms on both sides by soothing Argentine passions and allowing Peart to tell the British nation that moves were underway to reduce the risk of FMD importation. MAFF intended to send the mission in early February so that it could report back before the meat trade resumed in March; but on 6 February Peart received Reid's report on the origin of the epidemic. This concluded that lamb exported from Argentine *frigorifico* number 1408 was responsible for starting the epidemic. All action was put on hold as officials met to reconsider their options.[68]

Meanwhile, Peart's position became almost untenable as conflicting interest groups issued ever-more vocal demands. Politicians and farming organizations insisted that the ban must stay in place, at least until the independent committee of inquiry reported.[69] Though less well represented in the House of Commons, the National Federation of Meat Traders and a handful of consumers' representatives were equally adamant that it must be lifted. In Argentina, government officials and cattle producers proclaimed the ban an unjustified insult to their national honour. They continued to turn down British bids for public contracts, including the British Nuclear Export Executive's bid to build a UK£75 million nuclear power plant. Meanwhile, Argentine customs authorities intercepted British goods bound for the Falkland Islands.[70] Such reprisals were not publicized in Britain; therefore, many members of the public were mystified by Peart's continuing claim that the trade had to resume because the ban was only ever intended to be a temporary measure. Shropshire NFU complained: 'If there was no danger from imported meat, why was the ban imposed in the first place? If there is danger, why is it being lifted now? Clearly one of these decisions is wrong.'[71]

The government knew that it was in deep trouble. If it relaxed the ban and yet another FMD epidemic took hold, it would find itself in an impossible position. On the other hand, if the embargo remained *in situ*, Anglo–Argentine relations and the British economy could suffer further damage.[72] A Cabinet committee met to discuss the issue further. Peart pressed for the retention of the ban, and gained the support of the Commonwealth Office (which was motivated by the Irish fear of FMD spread) and the government's chief scientific adviser, Solly Zuckerman. However, the Foreign Office, supported by the Board of Trade and Treasury, insisted that it must be lifted because Argentine retaliation would cost the export trade an annual UK£25 million to £45 million, whereas FMD control would cost only UK£5 million a year. The divided committee then explored various compromise solutions. In the end, it selected a politically feasible yet scientifically nonsensical measure: the ban upon lamb and mutton imports would remain but beef imports would be admitted.[73]

Peart had the unenviable task of selling this policy to Parliament and the nation. On 4 March, he informed the House of Commons of the Reid report's findings, of his intention to resume imports of beef but not lamb, and of the forthcoming visit to Argentina of the British veterinary mission.[74] Critics cried 'shame!' and pointed out the illogicality of banning only sheep meat when cattle were equally susceptible to FMD. Sir A V Harvey MP saw straight through MAFF's window dressing and asked why a mission was needed when MAFF already had veterinary inspectors stationed in Argentina. But the real outcry came nine days later. By then, many members of parliament (MPs) had actually read the Reid report and found little in it to justify Peart's chosen course of action. The debate lasted nearly six hours; as member after member rose to denounce his partial lifting of the ban. Nevertheless, the measure passed, thanks to the government's large majority.[75]

Participants in the Argentine meat trade were extremely relieved to hear of the partial resumption of trade, but they were infuriated by the Reid report, which had implicated their lamb upon the basis of circumstantial evidence relating to the origins and destinations of imported meat cargoes. None of the suspected meat had been tested to see if it contained virus because, by the time its role was ascertained, most had been eaten.[76] British farmers and MAFF officials knew from experience that it was extremely rare to obtain definite evidence of disease origin and readily accepted the Reid report; but the lack of direct scientific proof enabled the Argentines to reject its conclusions.[77]

On 18 March, in the midst of this furore, the veterinary mission arrived in Buenos Aires. Members included Deputy CVO Albert Beynon and William Henderson, a former FMD researcher at Pirbright and past head of the Pan-American FMD bureau in Rio de Janeiro, who was already on amicable terms with the Argentine authorities.[78] Largely as a result of his presence, the mission achieved its goals, and in May 1968 Beynon presented its report to MAFF. It recommended several improvements to the Bledisloe agreement and a continued ban upon pork and mutton imports. It also suggested that MAFF halt offal imports and admit only boneless beef, since the tongues, kidneys, liver and bone marrow of infected animals contained the greatest concentrations of FMD virus.[79]

Although the mission's visit helped to restore Anglo–Argentine relations at a veterinary level, all was not well. In a propaganda move designed to demonstrate its rejection of the Reid report, the Argentine government refused to resume beef exports to Britain and continued to block the purchase of British exports.[80] Members of the British embassy in Buenos Aires, who already viewed the Argentines as an irrational and untrustworthy people, were astounded. Malcolm Gale, the commercial minister, remarked 'that the Argentine capacity for nose-cutting is almost unlimited', while Sir Michael Cresswell, the ambassador, attributed their reaction to a 'guilty conscience' and 'excessive emotion'.[81]

The Argentine government announced that it would not resume trade until the origin of the British epidemic had been properly investigated and demonstrated by 'the most rigorous scientific methods'. It then sent a mission to Britain, led by famous Nobel prize-winning physiologist and octogenarian Dr Bernard Houssay.[82] This concluded that FMD was endemic in Britain and that producer interests rather than scientific evidence had led to the ban on trade.[83] Satisfied, the Argentine government agreed to lift its embargo upon British exports; but trade resumed only slowly, much business was permanently lost and export licences were not granted freely until late July.[84] Meat exports to Britain also resumed, but under new conditions laid down by Garcia Mata, president of the Argentine National Meat Board. Instead of sending beef sides to Britain for sale by auction at a price that varied according to supply and demand, chilled beef was to be exported as partially boned, plastic-wrapped cuts, sold direct to supermarkets and restaurants, at a fixed price agreed in advance of shipment. It is not hard to guess Garcia Mata's motives. He must have known that British meat import policy was still under discussion, and probably hoped to persuade the British government that there was a viable alternative to an all-out trade ban. Moreover, the boneless trade was profitable because it used Argentine labour, allowed more efficient use of shipping space and resulted in a more predictable market.

British purchasers were not keen on the new system. They did not like to agree upon a fixed price for a highly perishable product several weeks in advance of its arrival, and consequently only 3500 tonnes of chilled meat left Buenos Aires between April and June 1968, 77,500 tonnes less than in 1966. Frustrated, new Minister of Agriculture Clewdyn Hughes urged retaliation against Argentina. Lord Brown, the minister of state, agreed that it was time we 'showed our teeth' and imposed trade sanctions. He complained that Argentina had behaved badly since the war, and that it was time the Foreign Office stopped pandering to its political games. However, Foreign Secretary Michael Stewart advised caution because sanctions would lead to an all-out trade war which would only benefit Britain's competitors.[85]

By October 1968, Anglo–Argentine relations were on the mend. Both governments signed a 'record of understanding', which stated that the origin of FMD outbreaks was not always capable of scientific proof, that MAFF would encourage Argentine participation in future investigations into British outbreaks, and that if meat were implicated, the Argentine government would be the first to know. In return, the Argentine government agreed to accept amendments to the Bledisloe agreement, which made FMD a notifiable disease, allowed only vaccinated meat to be exported, and laid down new controls upon vaccine safety and potency. Gradually, British purchasers began to accept the altered meat trade, and by January 1969, Argentine exports had returned to pre-ban levels.[86] But the crisis was not over, for the Northumberland committee of inquiry into the epidemic had not yet reported.

POST-MORTEM AND AFTERMATH

The Argentine reaction to the temporary meat import ban had made one thing clear to all British government departments, MAFF included. Regardless of the FMD risk and the wishes of British farmers, the importation of meat from South America simply had to continue. While it was feasible to adjust the conditions of trade, a renewed import ban was out of the question. It was important, therefore, to ensure that the Northumberland committee did not 'reignite Argentine passions' by recommending the latter policy. MAFF directed it to investigate only the 'technical' and not the political or commercial dimensions of FMD control in the hope that it would concentrate upon domestic control measures without becoming too heavily embroiled in the meat import issue. Senior officials also met several times with the Duke of Northumberland and informed him of the broader issues surrounding FMD control.[87]

Nevertheless, many officials continued to fear that the committee would recommend a complete import ban. They were especially concerned by one piece of evidence, a cost–benefit analysis of FMD control, prepared at the committee's request by A H Power and S Harris of MAFF's economic division. It predicted the likely number and impact of outbreaks during the period of 1969–1985 and quantified the direct costs involved in their control by vaccination, slaughter and a meat import ban, and slaughter together with a ban upon bone-in meat imports. The authors concluded that the second option would be the cheapest and most effective method of controlling FMD. Admitting only boneless meat would be equally effective but more expensive. Vaccination would be more expensive still, although if non-quantifiable effects were taken into account the difference between the policies was less marked.[88] Officials both within and outside MAFF feared that, on consideration of this data, the committee would immediately choose an import ban and therefore tried to prevent its submission in evidence. However, Carnochan insisted that it must go forward, partly because it provided a 'sober antidote' to the NFU's far larger estimate of the benefits deriving from an import ban, but also because it shone a favourable light upon a boneless beef import policy. This move had already been proposed by the veterinary mission and, in the light of recent trade adjustments, would probably prove acceptable to the Argentines. In agreeing to support Carnochan's bid, Tame commented: 'It is nice to be able to see a possible fall-back position emerging at this early stage!'[89]

In spring 1969, in anticipation of the first part of the Northumberland committee's report, British farmers agitated once more for a permanent meat import ban, tension rose in Argentina and the Foreign Office steeled itself for a renewed crisis in Anglo–Argentine relations.[90] In the event, the committee suggested a particularly diplomatic solution to the problem of FMD control. Instead of recommending a single policy, it presented several different options. Significantly, these connected the methods used to control FMD within Britain to the means by which FMD could be kept out of Britain:

> *The adoption of a policy which relies on the slaughter policy alone should, in our view, be dependent either on a complete ban on imports, or at least on the exclusion of the dangerous components of meat from countries or areas of countries where FMD is endemic. If these dangerous components are not excluded, we think it essential that some form of vaccination should be introduced.*[91]

A complete ban upon meat imports was the best policy 'on animal health grounds'. If, however, 'for social, political or commercial reasons' it proved unacceptable, the next best policy was to limit imports to boned beef.

In past epidemics, it had been MAFF's job to respond to such reports; but this time it became a matter for the 'Hughes group', an inter-departmental working party that sat to consider the political and economic implications of the report and to frame the government's response.[92] Members were quick to reject a complete trade ban and agree upon a boned beef import policy, subject to a substantial reduction in the existing 20 per cent tariff on this product. They expressed impatience with the Argentine government's political games and warned that if it retaliated – which it did not, despite veiled threats – they would impose a complete import ban.[93]

The Hughes group went on to consider the Northumberland committee's views on vaccination. The committee had decided that the technical
problems and the expense of vaccination meant that slaughter was still
preferable to the routine prophylactic immunization of all susceptible
British livestock. However, it felt that because modern vaccines were much
cheaper, safer and more effective than when the Gowers committee
reported, ring vaccination was a feasible and, indeed, valuable method of
FMD control, especially as the growth of intensive farming meant that
future epidemics might be even larger that that of 1967–1968. If meat
import policy failed to substantially reduce the risk of FMD invasion, then
'the slaughter policy should be reinforced by a ring vaccination scheme'.
In any case, 'contingency plans for the application of ring vaccination
should be kept in constant readiness'. One member, Anthony Cripps, went
so far as to issue a minority report, stating that the risk of FMD invasion
and spread meant that 'The immediate application of ring vaccination to
any outbreak which occurs seems to me essential.'[94]

The idea of emergency ring vaccination appealed to the Hughes group
and to the Cabinet. Therefore, having persuaded a reluctant Treasury to
part with the estimated UK£1 million a year needed for vaccine purchase
and storage, MAFF started to draw up contingency plans.[95] However, as
time went on and FMD did not return, its institutional resistance to
vaccination began to reassert itself. In September 1969, MAFF Under-
Secretary J Carnochan raised the possibility of giving 'Ministers and our
senior colleagues a further chance to consider whether we should go for ring
vaccination at all?' He claimed that the decision to press ahead with
contingency planning had been taken in a 'traumatic atmosphere'; but now
that the crisis had passed and the boned beef import policy had reduced
the FMD risk to a minimum, vaccination was probably not required.[96]
Reid's reply is extremely revealing and caused Carnochan to put aside his
reservations:

> *I have no heart for being the CVO who first pressed the button
> to push FMD vaccine into British stock; but against this I would
> not let my reluctance outweigh my judgement if circumstances
> arise which might light a fire that could not be put out without
> using every known defensive method of control. It would be
> irresponsible, following our experience in 1967–1968, not to use
> vaccine and to use it quickly if one saw a potentially dangerous
> situation.*[97]

Officials agreed that vaccination should remain a 'reserve' measure for use
in emergencies. It should be adopted only on the basis of veterinary advice

and without reference to interest groups outside the MAFF (a recommendation ignored in 2001, when the NFU was allowed to veto plans). Farmers should not be allowed to vaccinate at will because 'we must avoid undermining confidence in the slaughter policy and its status as the main line of defence'. The first-ever decision to vaccinate would be 'an historic affair, with emotional undercurrents, which could be the subject of controversy for at least a decade, and as such it would tend to be of a political/technical character'. Subsequent decisions, 'which, being based on actual experience and with the farming and general public acclimatized, might well be taken with less difficulty (and perhaps more quickly) and might come to be viewed primarily as a technical exercise'.[98]

Inevitably, when consulted on these plans, virtually all farming groups expressed firm resistance. Representatives argued that vaccination would hinder livestock exports, 'convey to the world that FMD is endemic in Great Britain', and 'change the standing of this country, which, in the past, we have been proud to call the stud farm of the world'. Moreover, for fear of acquiring 'carriers' of the FMD virus, farmers would not want to buy vaccinated stock. In direct contradiction to many of its earlier statements, MAFF claimed that such fears were unfounded. It also explained that slaughter would remain its main policy, but that it had to consider vaccination in order to avoid future criticism. It added a clause to the 1970 Agricultural Bill to enable it to enforce vaccination where necessary, maintained a vaccine bank of 1.5 million doses and instituted staff training exercises. The bank remained in existence until 1985, when for cost reasons it was replaced with an international vaccine bank.[99]

So, what can we deduce about MAFF's changing attitudes towards FMD vaccination? 'Shaken but not stirred' is probably the phrase that best describes the post-1967 state of affairs. In one sense, the crisis had brought about a momentous change, as officials broke with the past and acknowledged that there were circumstances other than biological warfare in which vaccination was required. They realized for the first time that under modern farming conditions, FMD might spread too quickly to be contained by slaughter alone. They also knew that public opinion, sickened by the recent death toll, would be less tolerant of slaughter in future. But their historic ties to slaughter and their professional, cultural and nationalistic antipathy to vaccination ran deep. They wanted to uphold Britain's status as a non-vaccinating nation, to maintain MAFF's record of defeating FMD by slaughter alone, and to stamp out germs without having to take on board the uncertainties of an unfamiliar technology. The world might have changed, but vaccination had not. It was still a second-best, risky and costly policy, and they hoped against hope that they would never have to put their contingency plans into practice.

For the Argentine government, the crisis had demonstrated the benefits of a retaliatory foreign policy. Had it quietly accepted responsibility for the epidemic and recognized the validity of the trade embargo, then FMD would have remained a purely agricultural problem, a matter for British farmers and MAFF officials who, given Britain's reduced dependence upon Argentine meat imports, would have had little difficulty in ensuring that the temporary meat import ban remained in place. As it was, Argentina's vocal criticisms of the meat ban and its refusal to purchase British goods at a time of economic crisis 'manufactured' FMD into a matter for the Foreign Office, Board of Trade and the Treasury. The veterinary rationale that had justified the trade ban was overthrown by pressing political and commercial issues, and potentially diseased meat once more found its way into Britain – but not for long. Argentine meat imports dropped off when Britain joined the EEC in 1973, and the following year they were banned, this time indefinitely and from the whole EEC. The reason was not FMD, but a serious glut of beef on the European market. Argentina could hardly carry out reprisals against all member nations, and so the ban stayed in place until 1977. By then, Britain's EEC ties prevented it from accepting extensive Argentine exports, and the 'special relationship' that had for so many years put British livestock at risk of FMD finally ended.[100]

For the British nation as a whole, the 1967–1968 FMD epidemic went down in history as the most brutal and devastating of all. The cost of disease control and the resulting death toll were unprecedented, and the social, economic and psychological impact of the disaster was felt far beyond the farmyard. Rural communities, businesses connected with agriculture, participants in the meat and livestock trade, veterinarians, city dwellers who engaged in countryside leisure pursuits, officials of all government departments, exporters to Argentina: all were touched by FMD. However, once stamped out, FMD did not return to Britain. Businesses recovered, farming picked up and the epidemic passed into memory. But the disease still lurked abroad, and it was only a matter of time before it returned to wreak havoc on an unforeseen scale.

Chapter 8

Foot and Mouth Disease, 2001

FIGHTING FMD, 1968–2000

As previous chapters showed, between 1886 and 1968, foot and mouth disease (FMD) was a fact of life in Britain, as barely a year went by without numerous fresh invasions of virus. But with the stamping out of the great 1967–1968 epidemic, the nation entered a long period of disease freedom. As the years passed and FMD did not return, farmers, veterinary surgeons and officials of the Ministry of Agriculture, Fisheries and Food (MAFF) began to nurture hopes that the recurrent scourge had at long last been defeated. They received a scare in 1981, when FMD appeared in the Isle of Wight and caused 13 outbreaks. But the disease did not spread to the mainland, which by the turn of the 21st century could boast over 30 years of freedom from FMD. So, was the absence of FMD fortuitous? Or did it represent the final victory of a man-made disease-control policy, born in Britain during the later 19th century and exported throughout the developed world? And how and why did this period of FMD freedom end, in 2001, with one of the most devastating epidemics ever experienced?

The later 20th century saw a marked improvement in the international FMD situation, which contributed to Britain's long period of FMD freedom. As Chapter 5 revealed, following the 1951–1952 pan-European epidemic, Ministry of Agriculture and Fisheries (MAF) officials and former Chief Veterinary Officer (CVO) Thomas Dalling led the formation of a European body, the European Commission for the Control of Foot and Mouth Disease (EUFMD), and formulated a systematic, international campaign against the disease. The EUFMD was slow to get off the ground because the French government opposed its establishment and tried to constitute a rival body under the Paris-based organization for international animal disease control, the Office Internationale des Epizooties (OIE, established in 1924).[1] But by the 1960s, the EUFMD's membership was increasing and it was beginning to have a real impact upon the European FMD situation.[2]

At the same time, European governments vigorously attacked FMD spread using a mixture of slaughter and vaccination.[3] In former years, there had been a shortage of vaccines, and most nations had had to use their supplies strategically, immunizing animals along national frontiers or in areas surrounding disease outbreaks. But the early 1960s saw the development of new vaccine production techniques at Pirbright. Scientists discovered that instead of having to culture the virus needed for vaccine production in tongue tissue obtained from slaughterhouses (the 'Frenkel' method, devised by Dutch scientist H S Frenkel in 1947), they could use huge vats of hamster kidney cells to produce virus continuously and in large quantities. Various commercial companies – dominated by the pharmaceutical company Wellcome and Co – took advantage of this new technology and erected chains of vaccine-production units across the world.[4] These developments increased the availability and reduced the cost of vaccines, allowing governments to regularly immunize all susceptible livestock within their borders. As a result, European FMD incidence dropped considerably, from 22,500 cases in 1960 to 3658 cases in 1968.[5]

However, some problems remained. It was still difficult to immunize species other than cattle, as illustrated by the widespread involvement of pigs in European outbreaks during the 1960s. This obstacle was later overcome by the development of new oil-based vaccines.[6] Then there was the threat that exotic viral types (primarily South African type 1, which was immunologically distinct from the A, O and C types more commonly found in the West) would spread from Turkey, where they were endemic, into Greece and Bulgaria.[7] At great cost, the EUFMD, in conjunction with the United Nations Food and Agriculture Organization (FAO) and the OIE, arranged for susceptible animals in the region to be vaccinated repeatedly against the prevailing virus types.[8] This *cordon sanitaire* was maintained from the mid 1960s until 1989. European FMD control was also hindered by the unknown disease situation beyond the Iron Curtain, and by the refusal of certain nations to join the EUFMD, notably France (which still maintained that the EUFMD duplicated the activities of the OIE), Germany (which acted out of solidarity with France) and Spain (for political reasons).[9] In 1974, Germany agreed to become a member, and following General Franco's death in 1975, Spain signalled its willingness to join. But although France regularly sent observers to the EUFMD meetings, it did not subscribe fully until 1981.[10]

By the mid 1970s, most European nations had eliminated endemic FMD, although they continued to suffer from sporadic outbreaks. At that time, most national governments laid down their own meat and livestock import policies; but France, Germany, Italy, Holland, Belgium and Luxembourg, which had joined together in 1957 to form the European Economic Community (EEC), adopted a common set of measures. Imported livestock had to originate from FMD-free areas and be vaccinated before entry, and member countries could, if they wished, ban meat imports from FMD-infected territories or nations.[11] Questions began to be raised about this policy in 1973, when Britain, Denmark and Ireland joined the EEC. While existing members all employed some form of FMD vaccination, these three nations used a slaughter policy and prohibited the importation of vaccinated livestock for fear that they could carry or mask FMD infection. Ireland refused to accept any meat from FMD-infected nations, whereas Britain would only permit imports from European regions situated over 20 kilometres from disease outbreaks, and reserved the right to stop the trade completely should widespread outbreaks or uncontrollable epidemics occur.[12] Both were reluctant to adopt EEC regulations on account of their valuable export trade with FMD-free nations such as Canada, Australia, New Zealand and the US, which restricted the importation of meat and livestock from infected or vaccinating countries. Existing EEC members therefore agreed upon special derogations that allowed Britain, Ireland and Denmark to maintain their existing FMD control policies.[13]

By 1980, endemic FMD had been mostly eliminated from Europe, although sporadic outbreaks continued to occur, many associated with poorly inactivated vaccines and escapes of virus from vaccine-production plants. These new circumstances led several European agriculturalists to demand an end to FMD vaccination in the belief that this would facilitate trade with non-vaccinating countries. Initially, the EUFMD rejected these calls. It claimed that Europe was still at risk from FMD invasions. Also, its intensive agricultural practices and large-scale livestock movements meant that, in the absence of vaccination, a widespread epizootic could easily take hold.[14] Before long, however, it became necessary to streamline European FMD control policies, because they were hindering the EEC's drive to remove trade barriers and turn the 'common market' into a genuine single market around which goods, services, people and capital could freely circulate. In anticipation of this transition to a European Union (EU, scheduled to operate from 1993), the EUFMD reconsidered its stance. It concluded that the economic justification for continuing mass vaccination was questionable and that member nations should abandon it in favour of a stamping-out policy.

However, some countries continued to support vaccination; therefore, in 1989 the EEC carried out a cost–benefit survey of FMD control. This estimated the likely numbers of primary and secondary outbreaks during the period of 1993–2003, and predicted a worst-case scenario of

20 invasions of FMD into the region, each of which could give rise to 150 secondary outbreaks. It concluded that under such circumstances, a Europe-wide slaughter policy would prove much cheaper than compulsory mass prophylactic vaccination. The Council of Ministers then decided that, from 1992, all EU members should stop vaccinating against FMD, adopt a compulsory slaughter policy and ban the importation of vaccinated livestock. Ring vaccination in the areas surrounding disease outbreaks would remain an option; but nations wishing to take this step would have to apply for EU permission. Governments should also formulate and submit contingency plans for the control of up to ten cases of FMD at any one time; MAFF's plan was approved in 1993.[15]

MAFF was extremely satisfied with the change in European FMD control policy. For almost a century, its officials had tried repeatedly to persuade other countries to adopt an ideal of national FMD freedom and British-style disease controls. As we have seen, they were moderately successful. This was partly because, as a net importer of meat and livestock, Britain could impose trade restrictions upon foreign infected nations. Also, thanks to Britain's international influence, had enabled MAF officials to establish the EUFMD and its four-tier system of FMD control, which framed vaccination as a necessary yet temporary step along the road to a slaughter policy. With the EU's decision to abandon vaccination, MAFF at last saw its policy preferences enshrined in international law. The impact of this move was felt far beyond the EU because countries wishing to export to the region also had to adopt a goal of national FMD freedom, favour slaughter over vaccination and restrict the importation of goods from infected or vaccinating nations. In this manner, MAFF's formerly contentious view of FMD became incontrovertible fact, accepted by all 'civilized' nations. At the same time, its disputed method of FMD control became set in stone, the basis for an international system of trade.

With the revamping of EU FMD control policy, the OIE laid down new guidelines upon how FMD should affect the international movement of meat and livestock. It set out a formal procedure for defining nations as 'FMD free', 'FMD free with vaccination' and 'FMD free without vaccination'. It also introduced new rules, which stated that when FMD-free nations succumbed to infection, they would not regain their former status – and would, therefore, be subject to the trade barriers of other FMD-free nations – until 3 months after the last case was stamped out, or 12 months after emergency vaccination ceased.[16] These measures both reflected and reinforced MAFF's belief that vaccination was risky and should only be used as a last resort, and they discouraged exporting nations from vaccinating even in 'emergency situations'. This was a paradoxical state of affairs. At a time when technological advances meant that vaccination was capable of controlling FMD more safely and more effectively than ever before, international opinion swung in favour of FMD control by slaughter.

Meanwhile, the South American FMD situation was slowly improving, although it still lagged far behind that of Europe. International cooperation played an important role, and was primarily mediated by the Pan-American FMD Bureau (Panaftosa). This body was established in 1951 at the request of the Organization of American States. It was based in Rio de Janeiro and worked under the auspices of the Pan-American Health Organization (PAHO), the regional office of the World Health Organization (WHO). Between 1957 and 1966 it was headed by British veterinary surgeon Dr W Henderson, who during the 18 years prior to his appointment had carried out much valuable research into FMD vaccines at Pirbright.[17] Panaftosa initially functioned as a research centre, examining methods of improving vaccination techniques and training personnel to participate in the control schemes of member countries. However, it later expanded its role to formulate plans for FMD control in South and Central America, and, during the 1960s, helped nations to develop bilateral FMD control agreements that required vaccination, inspection and the exchange of information upon disease incidence and spread. In 1972, it became the home of COSALFA (Comisión Sudamericana para la Lucha Contra La Fiebre Aftosa), a new body that, in its aims and status, was equivalent to the EUFMD.[18]

Progress within the region was slow owing to the size of South American nations, their political instability and their large populations of extensively farmed livestock. The poor-quality vaccines produced by many local companies hindered the efficacy of vaccination programmes. But, gradually, FMD incidence diminished. During the later 1980s, several areas were cleared of endemic infection and Chile was declared FMD free. The PAHO, working through Panaftosa, subsequently developed a hemispheric plan that set a target date of 2009 for the complete eradication of FMD from the region. The impetus to open up new markets in FMD-free areas of the world such as the EU, US and Japan encouraged farmers to cooperate. Aided by an improved disease surveillance system and new oil-based vaccines that produced longer, more reliable immunity, South American nations made substantial progress. Chile remained FMD free and, during the 1990s, Argentina, Uruguay and Brazil were declared 'FMD free with vaccination'. The former two countries later became 'FMD free without vaccination'. But in 2000, all three were hit by FMD. They lost their coveted 'FMD-free' status and their export markets, and Argentina and Uruguay were forced to restart vaccination programmes.[19]

Following the cessation of European FMD vaccination, sporadic disease outbreaks were stamped out in Italy (1993) and Greece (1994 and

1996). Several Eastern European nations that had stopped vaccination in order to trade with the EU also suffered occasional invasions: Bulgaria experienced FMD in 1991, 1993 and 1996; Russia in 1995; and Macedonia in 1996; while Turkey was subject to repeated outbreaks. In several cases, vaccination was temporarily applied and disease rapidly eliminated.[20] Fears of disease invasion and spread rose towards the end of the 1990s. In a 1999 paper, Y Leforban, EUFMD secretary, noted that the illegal importation of meat from FMD-infected regions such the Middle East, sub-Saharan Africa and parts of South-East Asia was an important route by which virus could enter Europe. He emphasized that if the virus did manage to invade, it was likely to spread faster and further than ever before. This was because the trend towards intensive farming had led to larger, more densely stocked farms, while the decline of international trade barriers had facilitated long-distance animal trading. He argued that European countries should reconsider their FMD control policies in the light of these risks, and ensure that they had adequate measures in place to prevent FMD invasion and to diagnose and control it promptly. They should also plan for the 'worst-case scenario' by preparing contingency plans for vaccination and carrying out simulation exercises.[21]

In March 1999, the European Commission's Scientific Committee on Animal Health and Animal Welfare issued a 'Strategy for Emergency Vaccination against FMD'. This viewed vaccination as a viable supplement to FMD control by slaughter. It argued that recent scientific advances had made vaccines extremely potent, and that their use would counteract the problems of manpower shortages, carcass disposal and public opposition that were commonly associated with a slaughter policy. It also pointed out that the development of new tests to differentiate vaccinated from infected animals (a previously impossible task since both possessed antibodies to FMD) meant that, in future, vaccination need not necessarily hinder international trade. The report suggested several vaccination strategies and listed considerations relevant to their adoption, such as the potential for large-scale movements of infected animals and the distribution and incidence of FMD.[22] Meanwhile, fears of FMD invasion increased following reports that a pan-Asia O strain of virus had spread beyond its traditional 'endemic' areas to infect Japan, Korea, Mongolia and South Africa, some of which had not experienced FMD for decades. European experts, meeting in September 2000, decided that within the next five years FMD was highly likely to invade Britain or Scandinavia.[23]

Therefore, at the turn of the 21st century, after a long period of improvement in the international FMD situation, many nations found themselves increasingly at risk of viral invasion. Significantly, late 20th-century developments in agriculture and commerce meant that any invasions of FMD were likely to cause greater economic and animal welfare problems than ever before. As Leforban had noted, farms had become larger and more densely stocked, so any single FMD outbreak was likely to involve more animals than during past years. The shift towards intensive farming had favoured the selective breeding of highly productive livestock, which would suffer more severely from FMD infection than 'unimproved' or indigenous stock. Meanwhile, under the influence of the EU and World Trade Organization (WTO), international trade barriers had declined, allowing European nations to increase their livestock exports. This valuable trade would be hard hit by the appearance of FMD.

These developments meant that, by 2001, FMD was no longer the same disease that it had been in 1968. Changing commercial, agricultural and political circumstances had 'manufactured' it into a more terrible plague than ever before, with an increased capacity to invade, spread, harm animals and inflict economic damage. This new state of affairs called for a re-evaluation of the methods used to prevent and control the disease. But although members of the British State Veterinary Service – including the CVO – thought that their 1993 contingency plan for FMD control needed updating, MAFF did not regard this as a priority. After all, in BSE and swine fever, officials had far more recent and pressing disease problems to deal with.[24] They apparently saw little point in devoting time to a disease that had not been seen on the mainland for over 30 years, and thought that its virtual disappearance from northern European and South American countries made its reappearance in Britain extremely unlikely. They were also confident that if FMD did invade, it could be rapidly stamped out by means of slaughter. They therefore ignored developments in vaccine technology and took little heed of the global resurgence of FMD. Consequently, the reappearance of FMD in February 2001 found MAFF unprepared, lacking in knowledge and inexperienced in FMD control, with totally devastating consequences for the nation.

EPIDEMIC, 2001

In February 2001, the official veterinary surgeon at Cheales abattoir in Essex noticed that three groups of pigs awaiting slaughter were extremely lame. Although he had never seen FMD before, he began to suspect the disease after finding blisters on their feet. A visiting MAFF veterinary officer agreed with his diagnosis and sent tissue samples to Pirbright for testing. Positive results were obtained and on 20 February, FMD was confirmed in Britain for only the second time since 1968.[25]

News of FMD led to an immediate ban upon the export of British meat, livestock and various agricultural products. As shock waves rippled through the British farming community, MAFF staff began tracing the origin and spread of disease. It transpired that the affected pigs had begun their journey far away, at Burnside farm in Northumberland. Although the route by which they had contracted infection was never definitively traced, officials deduced that they had been fed inadequately boiled swill containing infected meat, a practice connected with countless past FMD epidemics. The question was: where had this meat come from? Imports from regions where FMD was endemic were forbidden, and vaccinating nations were permitted to send only de-boned beef, which was extremely unlikely to convey the virus. However, illegal meat smuggling from FMD-infected regions was an acknowledged problem and seemed to be the most likely cause of the epidemic. MAFF responded to this news by banning swill feeding (upon which only 1.4 per cent of British pigs now depended). It also prosecuted the owner of the infected pigs, Bobby Waugh. He was later found guilty of feeding unprocessed waste, causing unnecessary animal suffering and failing to notify the authorities of the presence of FMD.[26]

MAFF's hopes that this would be a small and easily contained FMD outbreak were soon dashed as officials discovered that disease had already spread from Waugh's farm to infect neighbouring sheep. These animals had since passed through Hexham market in Northumberland and Longtown market in Cumbria, potentially transferring infection to thousands of other animals. Three days after the discovery of FMD, MAFF imposed a nationwide standstill upon livestock movements; but it was all too late. During that period, thousands of infected animals had been moved around Britain.[27] So, like the 1922 epidemic, this was a 'market' infection in which rapid, extensive livestock movements sowed the 'seeds' of hundreds of outbreaks. Infection had spread as far as Devon, Durham, Hereford, Lancashire, Anglesey in Wales, and Dumfries and Galloway in Scotland.[28] It had also invaded Ireland, France and The Netherlands. So Britain was no longer the innocent victim of other countries' failure to control FMD; instead, it had become a perpetrator of the crime.

MAFF responded to the rising number of disease notifications by applying the traditional slaughter policy, as laid down in the 1993 contingency plan. But as in 1922–1924 (and, to a lesser extent, 1967), its resources were rapidly overwhelmed. There was a desperate shortage of vets as a result of 50 per cent cuts in the State Veterinary Service during the previous 20 years. The long absence of FMD meant that the machinery for its control was no longer well oiled. Staff were poorly organized and lacked experience and expertise, MAFF's database – which supposedly held details of all British farms – was both inaccurate and out of date, and official ignorance of the scale of sheep movements in Britain enhanced the inefficiency of disease tracing and control.

Another problem was the flood of tissue samples sent by veterinary surgeons in the field to the now-renamed Institute for Animal Health (IAH) at Pirbright for FMD diagnosis.[29] Tests took several days, and until positive results were obtained, infected animals remained alive and continued to manufacture virus. This problem had not arisen during 1967–1968 when the disease had affected mostly pigs and cows, species which displayed characteristic disease symptoms. In 2001 however, FMD affected thousands of sheep, which showed few, if any, clinical signs of infection. The tardy diagnosis of FMD and ever-increasing delays in the slaughter and disposal of infected animals generated vocal, widespread criticisms of MAFF's bureaucratic, over-centralized and poorly organized control policy. Farmers also grew angry at its public claims that FMD was under control when, from their perspective, the situation was worsening by the day.[30] However, in contrast to 1924, those who experienced delays in the slaughtering of infected animals did not claim that FMD was a mild and inconsequential illness. This was partly because of changing perceptions of animal welfare, but also because FMD inflicted greater injury upon their modern, selectively bred livestock than upon the less-improved animals that had once populated Cheshire's farms.

MAFF and the State Veterinary Service were initially reluctant to ask for assistance from the army, the local authorities or Whitehall; but by mid March, as disease incidence spiralled and the media grew increasingly critical, it became plain that drastic action was required if FMD was not to become endemic. MAFF therefore instructed its veterinary staff to stop waiting for the results of laboratory tests and to 'slaughter on suspicion' animals that were potentially infected. This marked a shift to the methods of FMD diagnosis applied to Irish livestock imports during 1912–1923 (see Chapter 2). Inexperienced veterinarians, who feared the consequences should they misdiagnose FMD, began to see the disease everywhere, and even ordered the slaughter of livestock that had no 'classical' FMD symptoms and no history of exposure to the disease. Furthermore, because they were not used to looking in animals' mouths, they often did not know what had caused the lesions they discovered; on the whole, however, they played safe and confirmed FMD.

As the situation worsened, MAFF began to contemplate a 'firebreak' cull in Scotland and Cumbria of all sheep within 3 kilometres of infected premises. Computer models of the likely course of the epidemic, devised by a team of epidemiologists, lent support to this move, when combined with the cull of all susceptible livestock on farms contiguous to infected premises. Although there were doubts about the legality of an extended cull, MAFF nevertheless adopted it and set targets for the slaughter of all animals on (suspected) infected premises within 24 hours of notification, and of animals on contiguous premises within 48 hours. Army personnel were drafted in to provide the necessary logistical skill. Meanwhile, rising awareness of the extent of the crisis led the prime minister to claim that he was taking personal control of the situation. By opening of the Cabinet Office Briefing Room (COBRA), he brought the 'full force of government' to bear upon the disease.

In the carnage that followed, animals died in their millions.[31] Some – a minority – were undoubtedly suffering from FMD. Others died because they showed symptoms that might have been caused by FMD and no one wanted to risk leaving them alive. But the vast majority were completely healthy; they just happened to be in the wrong place at the wrong time. Some were killed because, as cloven-footed animals living within 3 kilometres of an infected place, they might have become infected. Others died under the somewhat incongruously named Welfare Scheme, which MAFF introduced in the wake of its nation-wide ban on animal movements. This ban had trapped many livestock in unsuitable accommodation and without adequate food when the lambing season was about to start and when many grazing stock were due to move to spring pastures. As photographs of lambs drowning in mud began to appear on the nation's TV screens, MAFF stepped in to offer farmers compensation for culling suffering animals.[32]

The combined effect of these policies was devastating. The number of disease outbreaks occurring in 2001 was 2026. This was actually less than the 2228 outbreaks of 1967–1968 and the 2691 outbreaks of 1923–1924. But in 2001, over 10 million animals died, whereas the death tolls in the earlier epidemics were 442,000 and 300,000, respectively.[33] And the reason for this carnage lay not in the disease itself, but in the government's 'manufactured' response to it.

MAFF's conviction that pre-emptive culling was vital for FMD control received a mixed reception. As in the epidemics of 1922–1924 and 1951–1952, farming opinions were divided. The National Farmers' Union (NFU) was, by 2001, the most prominent and politically influential of several British farming organizations, and was generally supportive of MAFF's FMD control policy, mainly because it wanted to resume exports as quickly as possible. However, a significant proportion of farmers rejected the NFU's stance and demanded FMD vaccination. Among them were many organic farmers, those living within the worst affected areas of Devon, Cumbria, Wales and Scotland, and the owners of rare breeds or irreplaceable hefted sheep.

The veterinary profession was similarly split between guarded support for MAFF and fervent opposition to its FMD control methods. In a situation that recalled the 1924 dispute between veterinarians and doctors (see Chapter 5), many vets were angry that responsibility for FMD control policy had been transferred away from the State Veterinary Service and the IAH's scientific experts, and handed to a group of epidemiologists who had no prior experience of the disease.[34] Leading veterinarians were not alone in claiming that the epidemiologists had gained the ear of government as a result of undue political influence, and were driven not by a humanitarian concern for the FMD situation but by the desire for professional advancement. Critics argued that the computer models upon which the extended cull was based were statistically flawed, that they had failed to consider vaccination as a policy option, and that they had not taken the practicalities of slaughter into account.[35]

Some of the strongest criticisms of MAFF's actions came from the tourist industry, which suffered far more extensive damage in 2001 than during past years. MAFF insisted on closing the countryside, a move considered in 1967 but rejected as impractical and unhelpful in controlling disease spread. Its actions, together with the widespread film footage of burning carcasses disseminated by the world's media, caused a steep decline in income from tourism, especially within the worst affected areas of Devon and Cumbria. The tourist industry had grown substantially during the later 20th century and was of far greater economic importance than the agricultural export market, which had halved in size since 1967–1968. Consequently, the selection of a FMD control policy that prioritized the agricultural interest over tourism caused seething discontent.[36]

Opponents of the official FMD control policy were largely undisturbed by the impact of FMD upon Britain's agricultural export trade. However, they were intensely concerned about the threat that the disease posed to their local communities, economic solvency, way of life, personal and professional relationships, and emotional and psychological health. They did not agree with MAFF's claim that FMD could only be controlled by a universal, centralized policy of slaughter, and thought the price it demanded for disease elimination was too high. They complained – as they had in 1924 and 1952 – that culling was an unscientific, unethical and barbaric method of FMD control. They also felt – and loudly said – that the contiguous cull was unjustifiable on the grounds that vaccination, coupled with more discriminate slaughter, could have achieved the same national goal at a considerably lower personal cost.[37]

As the epidemic lengthened, MAFF's claim that slaughter represented the cheapest, most effective method of stamping out FMD became progressively weaker.[38] However, together with the NFU, it rejected the use of vaccination because under OIE rules this would have delayed the resumption of the export trade. In fact, from the very start of the epidemic, the information disseminated by these bodies (which the national media often repeated, unquestioningly) portrayed vaccination as a 'last resort'. Using virtually the same arguments as in 1952, it labelled this technology unsafe, ineffective and practically problematic, while claiming that it would mask FMD infection or induce a carrier state. It contrasted the 'unknown' nature of vaccination with the historically successful and familiar slaughter policy. It also alleged that Britain could learn nothing from vaccinating nations such as Uruguay and The Netherlands because their FMD situation, farming practices and livestock distribution were very different from in Britain. Significantly, as in past years, most of these reports were directed towards discrediting mass vaccination, whereas ring vaccination – the most feasible policy, which had been recommended by the 1968 Northumberland committee and was permitted under EU law – received little mention.

As in 1952, pro-vaccinators claimed that MAFF had exaggerated the deficiencies and risks of vaccination. They asked how vaccines – which had significantly contributed to human and animal disease control – could be dangerous, and presented historical examples of their effective use overseas. Prominent contributors to this debate included several foreign and British FMD experts, whose advice MAFF pointedly ignored. They detailed the various ways in which vaccines could assist British FMD control and countered MAFF's propaganda about the dangers involved. And, as in 1952, scientists at Pirbright did not escape criticism, as it transpired that prior to the 2001 epidemic they had failed to validate tests for the 'penside' diagnosis of FMD that could have considerably reduced the culling of healthy animals.[39]

It is difficult to escape the conclusion that national pride once again influenced MAFF's insistence upon maintaining the slaughter policy. It had never had to vaccinate, and it did not want to start now. It had showed the world how to control FMD effectively, and was not about to sacrifice its much-lauded principle of FMD freedom and adopt a policy that all civilized nations had abandoned.[40] But such pride was sadly out of place. Recent epidemics of BSE and swine fever had put paid to the British boast of producing the finest, healthiest livestock in the world, and few people at home or abroad thought that the slaughter of 10 million livestock – so vividly depicted by the world's media – was anything to be proud of. The application of this policy failed to convey the image of a rational, organized, educated Britain, and it diminished rather than enhanced Britain's international standing.

Other elements of the British government began to view vaccines more favourably as the disease situation grew increasingly desperate, and plans were drawn up for vaccination in selected areas. But in direct contradiction to the post-1967 recommendations of its veterinary staff, MAFF refused to act without consulting and gaining the support of the NFU, which doggedly resisted the move. As in the 1923–1924 Cheshire epidemic (Chapter 3), it overruled the wishes of farmers within the worst affected areas and insisted that slaughter continue. It claimed that FMD was too widespread for vaccination to work and that consumers – who, in the wake of BSE, were extremely suspicious about meat quality – would not believe that vaccinated meat was safe to eat.[41]

Pro-vaccinators were furious. They felt that only vaccination could halt the culling of healthy stock, preserve valuable and rare animals, and enable the rapid reopening of the countryside. They claimed that the arguments against vaccination were self-interested and flawed, and pointed out that consumers were already eating meat from animals that had been vaccinated against numerous other diseases.[42] But although discussions about vaccination continued in specific regions of the country, plans were never implemented.

For farmers and their families, the experience of FMD had changed little since 1967 and even since 1922–1924. Thousands lived through the tragedy of seeing their stock suffer, die and be burned. But then, at least for them, the anxiety was over and a compensation cheque was due. Many more were forced to wait out the epidemic, living every day in fear that the disease might strike. They withdrew into their homes, surrounded their farms with disinfectant baths and sealed off the entrances. British farming was already struggling in the face of foreign competition and in the wake of the BSE disaster. Now the crisis deepened. Unable to move their stock without going through a lengthy, complicated licensing procedure, farmers were denied the means of making a living. And they were not alone. Workers in the tourist industry, auctioneers, dealers, hauliers, shop owners and publicans in rural areas, and many, many more found their businesses curtailed and their lives on hold. But there was a new source of strength – the internet. Websites and newsgroups sprang up as people sought to make sense of events. Many issued complaints about MAFF's actions, comforted those who were suffering, organized pro-vaccination campaigns and publicized various outrages that MAFF staff had allegedly committed in the course of their duties.

Although the impact of FMD was felt most intensely in rural areas, city dwellers also found their activities curtailed as the countryside closed to sports and tourist activities. While few of them experienced the disease at first hand, intense media interest in the epidemic meant that none could avoid hearing about it. FMD was more newsworthy than ever before and dominated the national press for nearly a month. In contrast to past years, there was less factual reporting of FMD spread and more analysis of its implications. Many journalists and columnists were highly critical of

MAFF's actions and they used the epidemic to question related issues, such as the validity of intensive farming, the nature of government, the state of rural Britain, the role of science and scientists in policy formation, the political influence of the NFU and the ethics of the veterinary profession.[43]

As in earlier epidemics, the apparent failure of the slaughter policy precipitated a debate over who was to blame. Many farmers alleged that MAFF veterinary surgeons, slaughterers, labourers and hauliers had contributed to disease spread by carrying virus between premises upon inadequately disinfected vehicles and clothes, and by failing to slaughter and dispose of infected animals promptly. They also claimed, as they had in 1924, that virus spread in the smoke from funeral pyres. They attacked MAFF's lack of preparedness, its poor organization, its lack of heed for recent warnings about the increased FMD threat and its failure to abide by the Northumberland committee report. The latter had recommended prompt slaughter, calling in the army quickly, minimal restrictions upon the countryside and contingency planning for vaccination. Critics resented the fact that FMD was controlled by a universal policy that did not take account of local or individual circumstances or discriminate between different degrees of risk of infection. Some alleged that MAFF officials had acted illegally or immorally in their efforts to enforce the cull of healthy animals, and that they had intimidated farmers into permitting the slaughter, lied about their right to appeal and broken into private premises to kill stock. A significant number of livestock owners (200 in Devon alone) turned to lawyers and the courts to protect their animals from the contiguous cull.[44] Such challenges had occurred rarely during previous epidemics, when the legality of the slaughter policy had not been in doubt and when healthy animals on contiguous premises were rarely slaughtered.

MAFF and the NFU vigorously denied such criticisms and attempted, as during earlier epidemics, to divert blame for the ongoing spread of FMD towards those 'awkward' farmers and dealers who refused to back the official control policy. Officials still saw slaughter as a disciplining, moralizing policy under which all members of the population had to work together towards the common good. They argued that it was the only scientific, rational and historically proven method of FMD control, and that all intelligent and enlightened individuals supported it. Critics of this policy were ignorant, immoral and selfish. They did not understand the many reasons why vaccination would not work, failed to realize that MAFF was acting in the interests of the nation and, in contesting the cull of potentially infected animals, were responsible for prolonging the epidemic. MAFF also alleged that many farmers and dealers had assisted the spread of infection by failing to abide by government 'bio-security' recommendations. Some had deliberately infected their animals in order to claim government compensation; others had tried to cash in on the generous valuations offered by the welfare scheme. Such claims angered farmers, who viewed themselves as the victims rather than the perpetrators of the crisis. Instead of drumming up support for the slaughter, MAFF's pronouncements heightened popular mistrust and strengthened resistance to the extended cull.[45]

During April 2001, FMD incidence decreased and the epidemic ceased to be headline news. Heartened by the decline, MAFF decided, from the end of April, to allow more discretion in applying the extended cull policy, a move that cynics claimed was stimulated by media interest in Phoenix the calf, whose future hung in the balance after he was overlooked during a farmyard cull.[46] Phoenix lived, and as the extent of the slaughter declined, popular resistance gradually diminished. But disputes continued to rage as proponents of the extended cull, who claimed that their policy had evidently worked, clashed with its critics, who argued that vaccination or the traditional slaughter policy would have had a similar impact upon disease incidence and caused considerably less suffering.[47] In June 2001, MAFF was itself culled and its officials absorbed into a new government body, the Department of the Environment, Food and Rural Affairs (DEFRA).[48]

FMD outbreaks continued throughout the summer and did not disappear altogether until September.[49] Critics began to call for a public inquiry into the epidemic, in the hope that this would reveal the full extent of the government's ineptitude. But the government argued that a public inquiry would be too lengthy and expensive. Instead, it established three overlapping enquiries, a Policy Commission upon the Future of Farming and Food led by Sir Don Curry, a Lessons Learned Inquiry chaired by Dr Iain Anderson, and a Royal Society Inquiry led by Sir Brian Follett, which was to investigate scientific questions of disease prevention and control.[50] Opponents complained that the three inquiries would be neither impartial, nor useful and that the important questions would 'slip between the cracks'. Their several High Court attempts to force a public inquiry were unsuccessful.[51] However, a European Parliament public inquiry – which the British government tried and failed to block – did take place. Various other inquiries were instituted at a local level, including the Devon, Cumbria and Northumberland inquiries and a Royal Society of Edinburgh inquiry. All proved highly critical of MAFF's handling of the epidemic and recommended far-reaching changes to policy. The following conclusion discusses some of their more important findings, and sums up what we have learned about the past, present and future of FMD in Britain.

Conclusion: Foot and Mouth Disease in Britain, 1839–2001 – Lessons Learned?

The calamitous reappearance of foot and mouth disease (FMD) in 2001 brought to the fore old questions about the nature of the disease, its means of spread and the best methods of controlling it. While unprecedented in its intensity, the shift to a wholesale cull policy was a logical extension of the traditional, century-old method of FMD control. Hence, at the start of the 21st century, British FMD control was based upon the same principles as the measures introduced during the later 19th century, when Queen Victoria was empress of India and the germ theory of disease had yet to be accepted.[1]

How can we account for the persistence of this policy in the light of the controversies that have repeatedly surrounded it? As Chapter 1 revealed, the vision of FMD upon which it was based – that of a foreign, invading plague that spread in almost every way imaginable and caused significant losses in meat and milk production – first came to the fore in Britain during the later 19th century. I have argued that the political, economic, scientific and social circumstances that helped to create this vision were largely specific to that particular time and place (late Victorian Britain). So it is no surprise that individuals experiencing FMD at different times and places came to very different conclusions. Moreover, since much of the fear surrounding FMD was 'manufactured' by the social, psychological and economic impacts of the measures used to control it, individuals encountering different control policies naturally adopted very different understandings of the disease.

We have seen how, at regular intervals, in Britain and abroad, various farmers, veterinarians, politicians and participants in the meat and livestock trade challenged the opinions of the British agricultural authorities. They made claims that FMD was a mild ailment, that infected animals quickly recovered, and that certain routes of virus spread were more important than others, and they reached very different conclusions about how it should be controlled. We should not label as 'wrong' these differing ideas about FMD. They may not accord with present-day British expert opinion; but if we take account of the geographical, political, cultural, administrative, agricultural, national and commercial framework within which they arose, they make good sense. Whether the disease was 'severe' in 1890 depended upon what kinds of animals you were raising and for what markets. Whether vaccination was promising in 2001 depended, in part, upon the likely attitudes of consumers and foreign regulatory authorities.

As the historical context changed, so did FMD. It has never been a fixed biological entity, with its nature, mode of spread and means of control clearly defined by science. Scientists have frequently disagreed over such matters to the extent that most disputants were able to find scientific evidence in support of their claims. We have seen that FMD was not, and is not, simply an animal issue; the issues it raised always involved economic and political circumstances and calculations, and every epidemic had major economic, political, social and psychological impacts. Since the 1870s, FMD has repeatedly brought fear, anxiety, tragedy and sorrow to Britain; it has curtailed businesses and social lives; it has shaped international relations; and it has altered the bonds that link different groups of society together.

So, if the FMD control policy devised by the late 19th-century British government was not the only or even the most obvious method of controlling FMD, and if the contexts of FMD have changed radically over the ensuing century, how can we account for the continuance – indeed, the intensification – of the policy? I have suggested that the answer lies largely in the powers and actions of British agricultural and veterinary officials, who were always convinced of the benefits of national FMD freedom. Because of Britain's international economic and political influence, they could ignore foreign criticisms of their meat and livestock import policies. At home, they could reshape the policy-making process so as to exclude their critics and grant easy access to supportive, influential bodies such as the National Farmers' Union (NFU). Their propaganda campaign exaggerated the benefits of slaughter and the drawbacks of vaccination, and branded as immoral and selfish those individuals who demanded policy change. When it suited them, they could conceal potentially inflammatory information from the public domain. And at times of crisis, when the policy was at risk, they learned to dampen down criticisms by making subtle adjustments, for example, by agreeing to isolate limited numbers of Cheshire cattle in 1924 and to moderate the extended cull policy during

April 2001. But at no point did they suspend the principles on which the slaughter policy was founded.

From the late 19th century until 1968, FMD control in Britain was almost continually under review. Five committees of inquiry sat between 1922 and 1968, and internal appraisals of policy occurred far more frequently. As a result, the British agricultural authorities introduced additional import restrictions, they proposed initiatives for international FMD control such as the Bledisloe agreement and the European Commission for the Control of Foot and Mouth Disease (EUFMD), and they stimulated intensive scientific research into vaccines. They also made contingency plans for serum administration in case Britain came under biological attack, and for ring vaccination should it suffer an epidemic more extensive than that of 1967–1968. By such adjustments, they kept the Victorian FMD control policy up to date and in touch with a changing world. Theirs was not the only way of controlling FMD, and history could have been different had other measures been tried. But when viewed in the light of Britain's geography and patterns of trade, slaughter remained a rational response.

However, after FMD disappeared in 1968, officials of the Ministry of Agriculture, Fisheries and Food (MAFF) lost touch with the disease. As time went by, their perceptions of, and policy responses to, FMD became increasingly dated and eventually ceased to make sense. In the belief that import barriers and improved international FMD controls had abolished the risk of FMD invasion, they paid no heed to warnings about the rising disease threat. They did not recognize that the enhanced capacity for disease spread within the nation meant that slaughter might not succeed in future, and they neglected to consider how advances in vaccine technology might impact upon domestic FMD control. They also failed to realize that after 30 years of freedom from FMD, the machinery for disease control was somewhat rusty, while staff familiar with its workings had almost all retired.

In the aftermath of the 2001 tragedy, the reports issued by the Royal Society, the 'Lessons Learned' and European Parliament inquiries were all extremely critical of MAFF's preparations for, and handling of, the epidemic.[2] They queried the reliance upon the traditional control policy, arguing that, however stringent, Britain's import barriers could never provide complete security against the introduction of FMD virus. They went on to note that new opportunities for FMD spread within the country meant that a single invasion of disease could rapidly outstrip the resources available for FMD control by slaughter. While 2001 had shown that radical slaughter could still contain a widespread FMD epidemic, it had also brought to light the severe social and psychological ramifications of this policy and its declining economic justification.[3] Disease elimination had cost the British economy UK£8 billion. Compensation to farmers amounted to UK£1.3 billion; tourism and associated industries lost between UK£4.5 billion and £5.4 billion.[4] But exports of livestock and livestock products – the trade that the slaughter policy was engineered to preserve – were worth just UK£1.3 billion a year.[5]

All the inquiries went on to recommend that future British FMD control policy incorporate vaccination, a technology that had advanced considerably during recent years and was safer than ever before. According to the Royal Society report, there were no longer any 'technical, scientific, trade or cultural' barriers to vaccine use. It echoed both the Northumberland and Gowers committees in claiming that the risk of vaccinated animals becoming 'carriers' of FMD had been exaggerated, and recommended that, henceforth, vaccination should be regarded as a 'policy of choice' instead of a 'last resort'. It also suggested that scientists validate new tests to differentiate vaccinated animals from those which had recovered from FMD, and suggested that these would permit the dismantling of trade barriers against vaccinating nations. The European inquiry thought likewise, and – recognizing that recent advances made vaccination a much less risky business – the Office Internationale des Epizooties (OIE) revised its regulations, allowing nations to regain 'FMD-free status' six months instead of one year after vaccination ceased.

The reports not only attacked the pro-slaughter ideology that had underpinned MAFF's control policy for nearly a century, they also recommended institutional changes to lessen the Department of the Environment, Food and Rural Affairs' (DEFRA's) tight control over the field of contagious animal disease control. They argued that FMD could no longer be seen as a purely agricultural problem, best managed by agricultural and veterinary officials, because it affected a range of different stakeholders who should also influence policy-making. Furthermore, FMD control was too centralized; in future, policies should be devised in consultation with local authorities and should take greater account of local circumstances.[6]

Such recommendations are laudable; but, as noted in the 'Lessons Learned' report, many of the problems and difficulties encountered in 2001 had been met in previous epidemics and were recollected by many farmers, veterinary surgeons and rural inhabitants.[7] Yet, during the crisis, MAFF policy-makers behaved as if that epidemic, and the subsequent discussion of vaccination, had never happened. Few, if any, had read the Northumberland committee reports. Nor had they bothered to examine the countless files compiled by their predecessors in an attempt to identify and correct the problems encountered during 1967–1968. So it was that the mistakes of 1922–1924, 1952 and 1968 were tragically repeated in 2001.

The 'Lessons Learned' report went on to suggest that 'the UK government should take the lead in the international debate' to ensure that vaccination became an internationally acceptable method of FMD control and did not impede trade.[8] It failed to recognize that in past years, MAFF officials had played a major role in creating those trade barriers between FMD-free and FMD-vaccinating countries, thereby building a global culture predisposed to FMD control by slaughter. In 2001, British farmers found themselves at the 'sharp end' of the policies that their government representatives had fought so hard to introduce. Now, it seems, Britain will push *towards* vaccination.

The Royal Society inquiry recommended more scientific research into FMD in the belief that this would help to resolve many of the problems of FMD control.[9] This suggestion may be too sanguine, especially if a 'scientific view' excludes the wider issues that have always shaped debates and, indeed, the scientific research itself. Additional investigations will not necessarily provide definitive answers to the problems of FMD prevention and control. Much depends upon scientists' preconceptions, upon who is funding or directing their work, and upon its interpretation by DEFRA policy-makers. And, in any case, what constitutes the 'correct' science of FMD, and who should decide how it influences disease control policy? The biology is but one component of an economic and social problem and in 2001 the problem was more the lack of history than any lack of science. So what of the future?

In response to the inquiries' criticisms, DEFRA published a revised FMD control contingency plan in March 2003.[10] This set out the various factors upon which a decision to vaccinate would be based, and laid down the procedures necessary to prevent an epidemic like that of 2001 happening again. But some observers remain suspicious. They claim that the plan does not do enough to ensure that vaccination would be used in future, and object to the fact that it includes pre-emptive and 'firebreak' culling of animals not exposed to disease.[11]

Suspicions that DEFRA intends to reapply its 2001 policy in future outbreaks were first aroused in autumn 2001, when it introduced a new Animal Health Bill, proposing legal power to enforce an extended cull. The bill confirmed critics' suspicions about the illegality of its actions during the 2001 epidemic. It was denounced in the House of Lords and defeated; but the government's large majority in the House of Commons ensured its passage in 2002. In their recent analysis of these events, University of Cardiff law professors David Campbell and Robert Lee complained: 'The government, rather than review the flaws in its policy. . .is avoiding any lessons to be learned by purporting to give itself the legal power to repeat its mistakes. . .it is legislation which intentionally gives a power to panic.'[12]

So, while the 2001 epidemic has opened a window for change in Britain's century-old FMD control policy, it is not yet clear what will happen when FMD appears again, as it surely will. After all, contingency vaccination plans have been made before, but were never used. Will DEFRA follow the example of the now-defunct MAFF and stick stubbornly to a slaughter-only policy? Or will it make a new kind of history and adopt vaccination? And how will its actions be judged? Only time will tell whether Britain can finally break the mould of this manufactured plague.

Notes

INTRODUCTION

1 For example, see Anon (1965); F Brown (2003)

CHAPTER 1

FMD strikes

1 Anon (1965), pp135–6; I Pattison (1984), ch 1 and 2; J Fisher (1993a)
2 'The murrain', *The Times*, London, 31 October 1840, p6, col e
3 N Goddard (1988)
4 Anon (1840); W Sewell (1841)
5 'The murrain', *The Times*, London, 31 October 1840, p6, col e; ibid, 28 December 1840, p3 col d; H Keary (1848), p446; F Clater (1853), p141; R Perren (1978), ch 4
6 G Brown (1873), p441
7 H Keary (1848), p446; F Clater (1853), p140
8 G Fleming (1869)
9 J Howard (1886); F Smith (1933); Anon (1865), p265
10 F Fenner and E P Gibbs (1993), p415; D C Blood and O Radostits (1994), p967
11 D Taylor (1975); R Perren (1978), pp59–63; J Walton (1986)
12 J Homfray (1884), p15
13 M Pelling (1978), ch 1; C Hamlin (1992); J Pickstone (1992); M Worboys (2000), pp136–78
14 Anon (1840); 'The murrain', *The Times*, London, 31 October, p6, col e; W Sewell (1841); J B Simonds (1857) Evidence to Select Committee on the Sheep, etc Contagious Diseases Prevention Bill, 1857; J Giblett, 'Cattle disease', *The Times*, London, 24 November 1863, p5, col e; C, 'The diseases in cattle', *The Times*, London, 4 December 1863, p6, col f

15   J Howard (1886), pp1–3
16   The Analytical Sanitary Commission (1851); J Caird speech, *Hansard* [HC], 9 March 1864, vol 173, col 1750; Evidence by J Giblett and J Honk to Select Committee on the Cattle Diseases Prevention Bill, 1864; 'The cattle disease question', *Farmers Magazine*, vol 25, 1865, pp119–22, 217–20, 331–33; J Blackman (1975); D Taylor (1975), pp19–20; R Perren (1978), ch 2
17   J C McDonald (1951); K Maglen (2002)
18   J Broad (1983); J Fisher (2003), pp315–17
19   E Ackerknecht (1948); R Morris (1976); O MacDonagh (1977), ch 8; M Durey (1979); A Hardy (1993a); A Hardy (1993b), ch 9; C Hamlin (1994); P Baldwin (1999) pp2–36; K Maglen (2002)
20   E P Hennock (1998), pp54–55
21   Anon (1965), pp162–63; I Pattison (1990), pp35–44
22   W Drake, Evidence to Select Committee on the Sheep, etc Contagious Diseases Prevention Bill, 1857; J Giblett, J Honk and W Simmonds, Evidence to Select Committee on the Cattle Diseases Prevention Bill, 1864

John Gamgee and the diseased meat problem

23   R D'Arcy Thompson (1974)
24   F Accum (1820); Anon (1830); J Mitchell (1848)
25   P J Rowlinson (1982), pp63–66; J Burnett (1979), pp72–90
26   R J Richardson, Evidence to *Third Report from the Select Committee on the Adulteration of Food*, 1856
27   Bill to Consolidate and Amend Nuisances Removal and Diseases Prevention Acts, 1848 and 1849
28   Evidence by J B Simonds to Select Committee on the Sheep, etc Contagious Diseases Prevention Bill, 1857; A Friend of Rory O-Moo-oh, 'Cattle-Disease', *The Times*, 1 December 1863, p10 col e
29   E Headlam Greenhow (1857)
30   J Gamgee, 'Unwholesome meat', *Lancet*, vol II, 1860, pp 595–96; J Gamgee (1863); R Perren (1978), ch 2 and 4; J Fisher (1979–1980)
31   *Fifth Report of the Medical Officer of Health of the Privy Council*, 1863, pp19–31
32   F Smith (1933); S Hall (1962), pp48–50; Anon (1965), p265
33   For more information, see A Hardy (1999)
34   J Gamgee, 'Cattle disease in relation to the public health' and 'The system of inspection in relation to the traffic in diseased animals or their produce', clippings dated 1863, contained within J B Simonds Collection, RVC, London; J Gamgee, Evidence to *Report from the Select Committee on Cattle Diseases Prevention Bill*, 1864
35   'North of England Veterinary Association quarterly meeting', *Veterinarian*, vol 37, 1864, pp654–66
36   See disputes surrounding veterinary obstetrics and an outbreak of disease among foxhounds, *Veterinarian* 1860 and 1863, passim

37 'The smallpox in sheep', *Farmers' Magazine*, vol 22, 1862, pp355–56
38 E Holland, 'On disease in cattle', and 'Mr Holland's practical suggestions as to the prevention of disease', clippings dated 1863, J B Simonds Collection, RVC, London; Bill to Extend the Provision and Continue the Term of the Act of the 12th Year to Prevent Spreading of Contagious and Infectious Disorders Among Sheep and Cattle, 1863
39 Bill to Make Further Provisions for the Prevention of Infectious Diseases amongst Cattle, 1864
40 North British Agriculturalist Office, 1863, communication to J B Simonds, J B Simonds Collection, RVC, London; *Report from the Select Committee on Cattle Diseases Prevention*, 1864; J Fisher (1979–1980), pp51–53
41 'The farm', *Illustrated London News*, 18 June 1864, J B Simonds Collection, RVC, London
42 *Illustrated London News*, 25 June 1864, J B Simonds Collection, RVC, London; Falkirk and Ballinasloe were two important fairs where store stock changed hands
43 *Hansard* [HC], 15 July 1864, vol 176, col 1537–39; Captain O'Brien, 'The Sale and Transport of Cattle', *Mark Lane Express*, 26 December 1864, J B Simonds Collection, RVC, London

The cattle plague epidemic, 1865–1967

44 S Hall (1962); Anon (1965), pp13–21, 125–34; J Fisher (1980); M Worboys (1991); J Fisher (1993); T Romano (1997)
45 'Editorial', *Veterinarian,* vol 40, 1867, pp396–99; G Fleming (1880)
46 Bill to Consolidate, Amend and Make Perpetual the Acts for Preventing the Introduction or Spreading of Contagious or Infectious Diseases among Cattle and other Animals in Great Britain, 1868–1869
47 'Mouth and foot disease', *Veterinarian*, vol 42, 1869, pp751–2
48 For a summary of these debates, see *Report of the Select Committee on Contagious Diseases of Animals* (1873); H Jenkins (1873); *Report of the Select Committee on Cattle Plague and the Importation of Livestock* (1877); *Hansard* [HC], 24 June 1864, vol 241, col 134–98; ibid, 25 June 1864, vol 241, col 331–408; ibid, 27 June 1864, vol 241, col 500–75; ibid, 1 July 1864, vol 241, col 500–73; *Report of the Lords Select Committee on the Contagious Diseases of Animals Bill* (1878). For more information on the nature and geographical distribution of farming patterns, see J Fisher (1980), pp286–92; P Brassley (2000a); B A Holderness (2000). For Victorian politics, see T A Jenkins (1996). For information on nutrition and meat consumption, see E Smith (1863); J Burnett (1979); A Rabinach (1992); H Kamminga and A Cunningham (1995)

FMD becomes a plague

49 *Evening Standard,* London, 8 July 1869, J B Simonds Collection, RVC, London; L P Curtis (1968); H Ritvo (1987), pp45–81; J Fisher (1993a); T A Jenkins (1996), ch 5; J Fisher (2000)

50 'Mr Henry Chaplin MP on FMD', *Bells Weekly Messenger,* 21 January 1884, p6; J Brown (1987) pp5–59; P Perry (1973), ppxiv–xxiii

51 M Worboys (1991), pp314–17; M Worboys (2000), 'Introduction'; T Romano (2002), ch 5; J B Simonds, 'Introductory address to the RVC', *Veterinarian,* vol 48, 1875, p724. For opinions relating to the cause of FMD, see witness evidence published in *Report of the Select Committee on Contagious Diseases of Animals* (1873); *Report of the Select Committee on Cattle Plague and the Importation of Livestock* (1877); *Report of the Lords Select Committee on the Contagious Diseases of Animals Bill* (1878)

52 'Report of the Council', *Journal of the Royal Agricultural Society of England,* vol 37 1876, ppv–x; J Burdon-Sanderson (1877); W Duguid (1877). T Romano (2002) surveys 19th-century career opportunities for medical scientists.

53 'FMD Order, 12/69' *Veterinarian,* vol 43, 1870, pp59–60; *Hansard* [HC] 15 July 1872, vol 212, col 1126; ibid, 14 February 1873, vol 214, col 509 and 520; *Report of the Select Committee on Contagious Diseases of Animals* (1873)

54 'RASE on cattle disease', *Veterinarian,* vol 47, 1874, pp456–58; *Hansard* [HL], 10 February 1876, vol 227, col 668–70

55 *Report of the Select Committee on Cattle Plague and the Importation of Livestock* (1877); *Report of the Lords Select Committee on the Contagious Diseases of Animals Bill* (1878); Bill for Making Better Provision Respecting Contagious and Infectious Diseases of Cattle and other Animals (1878); ibid, as amended in committee (1878)

56 Bill to Amend the Contagious Diseases (Animals) Act (1878, 1884); Anon (1965); *Annual Report of the Agricultural Department* (1885)

57 *Return of Number of Farms Infected with Foot-and-Mouth Disease* (1870)

58 Witness evidence, *Report of the Select Committee on Contagious Diseases of Animals* (1873)

59 *Annual Report of the Agricultural Department* (1883); 'Professor Brown at the RASE Veterinary Committee', *Veterinarian,* vol 56, 1883, pp256–62

60 *Report of the Select Committee on Contagious Diseases of Animals* (1873); *Report of the Select Committee on Cattle Plague and the Importation of Livestock* (1877); *Report of the Lords Select Committee on the Contagious Diseases of Animals Bill* (1878); Parliamentary Debates on the 1878 Contagious Diseases of Animals Bill, *Hansard* [HC], 1878, passim

61 *Annual Reports of the Agricultural Department* (1880–1883); *Hansard* [HC], 22 March 1881, vol 259, col 1662–729; ibid, 16 April 1883, vol 278, col 172–93; ibid, 10 July 1883, vol 281, col 1020–87. See also *Veterinarian, Veterinary Journal* and *Bells Weekly Messenger,* 1883–1884, passim; A H H Matthews (1915), ch 2

62   *Hansard* [HC], 14 February 1884, vol 284, col 838–50; ibid, 19 February 1884, vol 284, col 1291–1300; ibid, 21 February 1884, vol 284, col 1528–39; ibid, 18 March 1884, vol 286, col 162–211

63   See descriptions of FMD outbreaks in *Annual Reports of the Agricultural Department*, and reports upon disease spread in *The Times*, London, 1885–1906. For an analysis of reactions to epidemic diseases, see C Rosenberg (1992).

64   E Whetham (1979)

65   *Annual Report of the Agricultural Department* (1884–1892)

CHAPTER 2

The Irish question

1   D Spring (1984), pp 33–34; G Williams and J Ramsden (1990), pp368–71; A O'Day (1998), ch 9–11

2   *Annual Report of Department of Agriculture and Technical Instruction for Ireland* (1914), p59; G H Collinge, T Dunlop Young and A P McDougall (1929), pp139–42; R Perren (1978), pp95–100; K Miller (1985), pp362–64, 380–402

3   Obituary, Stewart Stockman (1926)

4   Above paragraphs drawn from *Report on FMD in Ireland in the Year 1912* (1912); *Annual Report of Proceedings under the Diseases of Animals Acts* (1912); *Annual Report of Department of Agriculture and Technical Instruction for Ireland* (1912); 'FMD–Ireland', *The Times,* London, 1 July 1912, p8 col b; ibid, 2 July 1912, p8, col d; ibid, 8 August 1912, p11 col d

5   These men played a prominent role during late 19th- and early 20th-century battles for Irish land reform. See D Jones (1983) and P Bew (1987).

6   *Hansard* [HC], 5 July 1912, vol 40, col 1487–1567

7   H Jenkins (1873); Irish evidence to *Report of the Lords Select Committee on the Contagious Diseases of Animals Bill* (1878)

8   C Bathurst, 'FMD–Ireland', *The Times*, London, 3 July 1912, p10, col c

9   'Special article', ibid, 15 July 1912, p13 col e; C Bathurst, correspondence, ibid, 3 September 1912, p6 col a; ibid, 20 September 1912, p8 col c; ibid, 1 October, p12 col d; *Hansard* [HC], 5 July 1912, vol 40, col 1487–567; ibid, 18 October 1912, vol 42, col 1607–82; A O'Day (1998), pp219–20

10   *Annual Report of Department of Agriculture and Technical Instruction for Ireland* (1912)

11   *Hansard* [HC], 5 July 1912, vol 40, col 1487–567; ibid, 8 July, 1912, vol 40, col 1749–55; ibid, 18 October 1912, vol 42, col 1607–82; 'The Irish cattle trade', *The Times*, London, 6 September 1912, p10 col a; 'Party politics and cattle disease', ibid, 16 September 1912, p8 col a; E Kennedy, correspondence, ibid, 16 September 1912, p3 col e; 'Mr T W Russell on Irish cattle', ibid, 18 September 1912, p4 col d; 'Irish Deputation', ibid, 5 October 1912, p4 col f

12 'Special article – FMD', *The Times*, 15 July 1912, p13, col e; 'The Irish cattle trade', ibid, 6 September 1912, p10 col a; 'FMD', ibid, 9 September 1912, p4 col a; C Bathurst, correspondence, ibid, 20 September 1912, p8 col c; 'FMD', ibid, 7 October 1912, p3 col e; *Hansard* [HC], 18 October 1912, vol 42, col 1607–82; ibid, 8 February 1913, vol 48 col 366

13 'Special article – FMD', *The Times*, London, 15 July 1912, p13 col e

14 'FMD – outbreaks, Ireland', *The Times*, London, 8 August 1912, p8 col c; ibid, 2 September 1912, p3 col d; ibid 18 September 1912, p4 col c

15 *Hansard* [HC], 8 February 1913, vol 48, col 409

16 *Hansard* [HC], 5 July 1912, vol 40, col 1502; ibid 8 February 1913, vol 48, col 349–450; Major Head, correspondence to *The Times*, London, 11 September 1912, p8 col d; Lord Mayo, ibid, 28 September 1912, p10 col d

17 *Hansard* [HC], 5 July, 1912, vol 40, col 1567; ibid, 8 February 1913, vol 48, col 349–450; 'Party Politics and Cattle Disease', *The Times*, 16 September 1912, p8 col a

18 Report of Dillon's letter to *Freemason's Journal*, *The Times*, London, 26 September 1912, p4 col d

19 *Hansard* [HC], 8 February 1913, vol 48, col 375

20 'Nationalist support for stock owners', *The Times*, London, 24 September 1912, p4 col d; R Sanders letter to C Bathurst, ibid, 15 October 1912, p12 col f; *Hansard* [HC], 18 October 1912, vol 42, col 1621–30, 1660; ibid, 8 February 1913, vol 48, col 409–11

21 *Hansard* [HC], 18 October 1912, vol 42, col 1625–29, 1656–62; ibid, 8 February 1913, vol 48, col 362, 364

22 *Hansard* [HC], 18 October 1912, vol 42, col 1659

23 'FMD – Ireland', *The Times*, London, 9 July 1912, p12 col a; *Hansard* [HC], 8 July 1912, vol 40, col 1749–55; ibid, 18 October 1912, vol 42, col 1651; *Report on FMD in Ireland in the Year 1912* (1912), pp819–21

24 *Hansard* [HC], 7 August 1912, vol 41, col 3284–90; 'The Irish cattle trade', *The Times*, London, 31 August 1912, p6 col c; C Bathurst correspondence, ibid, 3 September, p6 col e; Chaplin correspondence, ibid, 6 September 1912, p10 col b; Pretyman correspondence, ibid, 17 September 1912, p6 col e; 'T W Russell's position', ibid, 24 September 1912, p4 col e; 'Special article', ibid, 7 October 1912, p3 col d; 'Deputations', ibid, 10 October 1912, p3 col a

25 'The Irish cattle trade', *The Times*, London, 6 September 1912, p10 col a; 'Nationalist support for stock owners', ibid, 24 September 1912, p4 col d; Dillon correspondence to *Freemason's Journal*, ibid, 26 September 1912, p8 col d; ibid, 3 October 1912, p11 col e; *Hansard* [HC] 8 February 1913, vol 48, col 399–405

26 D Spring (1984), pp34–38

27 *Report on FMD in Ireland in the year 1912* (1912), pp842–44; 'The Irish cattle trade', *The Times*, London 30 September 1912, p4 col e

28 'Deputations to Mr Runciman', *The Times*, London, 10 October 1912, p3 col a; C Bathurst correspondence, ibid, 10 October 1912, p3 col b; Chaplin correspondence, ibid, 29 October, p12, col d; 'FMD' ibid, 28 October 1912, p12 col c; *Hansard* [HC], 18 October 1912, vol 42, col 1607–82

29   *Report on FMD in Ireland in the year 1912* (1912), pp853–57; *Annual Report of Proceedings under the Diseases of Animals Acts* (1912); 'FMD', *The Times*, London, 2 October 1912, p4 col d; 'Deputations to Mr Runciman', *The Times*, London, 10 October 1912, p3 col a; *Hansard* [HC],18 October 1912, vol 42, col 1607–82; ibid, 8 February 1913, vol 48, col 349–447

Getting to grips with FMD

30   *Annual Report of Proceedings under the Diseases of Animals Acts* (1912–1923), passim

31   *Hansard* [HC], 23 February 1914, vol 58, col 1457

32   D Spring (1984), pp 33–34; G Williams and J Ramsden (1990), pp368–71; A O'Day (1998), ch 9–11

33   H Skinner, 'Passive Immunisation against FMD: a review', FMDRC CP 411 (1940?), Pirbright Archive; S Smith Hughes (1977), pp11–87; T van Helvoort (1994), pp190–94; H P Schmiedebach (1999)

34   *Annual report of Department of Agriculture and Technical Instruction for Ireland* (1912–1916); *Report of Proceedings at a Conference held at Birkenhead on the 26th of February 1914* (1914); *Annual Report of Proceedings under the Diseases of Animals Acts* (1912–1923); *Annual Report of Department of Agriculture and Technical Instruction for Ireland* (1912–1916)

35   *Annual Report of Proceedings under the Diseases of Animals Acts* (1908, 1911, 1912); *Report of the 1912 Departmental Committee on FMD* (1912), including Evidence, Appendices and Index

36   *Report on FMD in Ireland in the year 1912* (1912), p805

37   L P Curtis (1968); L P Curtis (1997); S Gilley (1978)

38   RAS Correspondence with Royal Dublin Society, 31 March to 5 May 1914, archives of Reading University Rural History Unit; *Report of Proceedings at a Conference held at Birkenhead on the 26th of February 1914* (1914); *Hansard* [HC], 23 Feburary 1914, vol 58, col 1439–94; ibid, 16 June 1916, vol 63, col 956–1030; ibid, 13 December 1915, vol 76, col 1788–90; ibid, 20 April 1921, vol 140, col 1886–87; *Annual Report of Proceedings under the Diseases of Animals Acts* (1914–1922); *Annual Report of Department of Agriculture and Technical Instruction for Ireland* (1914–1916)

39   Report and correspondence, 1916–1917, PRO MAF 35/158

40   *Annual Report of Proceedings under the Diseases of Animals Acts* (1921), pp30–33

41   Evidence to and *Report of 1922 Committee of Inquiry* (1922), PRO MAF 35/162 and MAF 35/165

42   *Report of the 1922 Departmental Committee on FMD* (1922), p24

43   Anon (1965), p393

CHAPTER 3

Overview

1 On the culture of cattle breeding, see 'Barons of Beef' in H Ritvo (1987)
2 The above section is drawn from *Annual Report of Proceedings under the Diseases of Animals Acts* (1920–1924); *Report of the 1922 Departmental Committee on FMD* (1922); *Report of the 1924 Departmental Committee on FMD* (1924); Evidence to the committees, contained within PRO MAF 35/159, MAF 35/160, MAF 35/162, MAF 35/165; *The Times*, London, January–March 1922 and September 1923–January 1924, passim; RAS council minutes, 12 December 1923, Rural History Centre, Reading University

The Cheshire experience, 1923–1924

3 *Crewe Chronicle* (hereafter *CC*), Chester, 19 January 1924, p8 col a
4 E Driver (1909); G Scard (1981), p65; D Taylor (1987), pp48–50
5 *Cheshire Observer* (hereafter *CO*), Chester, 22 September 1923, p8 col a; *CC,* 3 November 1923, p7 col b; Correspondence, PRO MAF 35/164
6 *CO*, Chester, October 1923, passim; *CC*, Chester, 10 November 1923, p7 col c–e
7 *CC,* Chester, 17 November 1923, p7, col d–e; ibid, 24 November 1923, p4 col h; *CO*, Chester, 1 December 1923, p5 col d–e
8 Witness evidence, PRO MAF 35/164, MAF 35/165, MAF 35/166
9 Cheshire Farmers' Union minutes, 3 December 1923, Cheshire County Records Office; *CC,* Chester, 8 December 1923, p7
10 A Matthews (1915); N Goddard (1988), pp1–3, 154–5. For more information on the NFU's historic relationship with MAF, see P Self and H Storing (1963); G Cox, P Lowe, and M Winter (1986); ibid (1991); M Smith (1990); J Brown (2000)
11 'FMD', *Daily Mail,* London, 17 December 1923, p9; NFU Meat and Livestock Committee minutes, 18 December 1923, Rural History Centre, Reading University; *CC,* Chester, 29 December 1923, p4 col f–h; 'The policy of slaughter', *The Times*, London, 2 January 1924, p7 col a; H German correspondence, ibid, 8 January 1924, p8, col c; 'Farmers' policy', ibid, 17 January 1924, p18, col a; German contribution to committee of inquiry, PRO MAF 35/162, pp200, 204, 220
12 Committee evidence, PRO MAF 35/168
13 *CC,* Chester, 5 January 1924, p8; 'Cattle plague', *The Times*, London, 8 January 1924, p9, col a
14 'FMD', *Daily Mail,* London, 17 December 1923, p9; 'Cheshire farmers' complaints', *The Times*, London, 5 January 1924, p12 col c; Duke of Westminster correspondence, ibid, 6 February 1924, p11 col e; Committee evidence, PRO MAF 35/165, pp69–70 and PRO MAF 35/166

15   Those animal healers who did not possess the RCVS diploma were, under the 1881 Veterinary Surgeon's Act, placed on a register of 'veterinary practitioners'; F Bullock (1930), p15

16   Evidence by Thomasson, Barker and Boyle, PRO MAF 35/166, *Report of the 1924 Departmental Committee on FMD* (1924), p33; 'Farmers' Policy', *The Times*, London, 22 January 1925, p18, col c; ibid, 12 August 1925, p9, col c; H Skinner (1989), p35

17   'FMD still raging', *CC*, Chester, 8 December 1923, p7; 'Stirring story of the cattle plague', *CC*, Chester, 15 December 1923, p7

18   ibid; 'Cheshire's Disappearing Herds', *CO*, Chester, 29 December 1912, p8

19   Committee evidence, PRO MAF 35/164

A county under siege

20   *CC* and *CO*, Chester, 15 December, 22 December and 29 December 1923, passim

21   'Animal disease spreads', *Daily Mail*, London, 14 December 1923, pp9–10

22   'Comment' and 'Medical officer on cremation', *CC*, Chester, 22 December 1923, p7; *CO*, Chester, 29 December 1923, p8; J Crowe evidence, PRO MAF 35/164

23   Stockman evidence, PRO MAF 35/165, day 2, p48–50

24   RAS council minutes, 12 December 1923, Archives of Reading University Rural History Unit; 'FMD', *The Times*, London, 14 December 1923, p4, col e; *CC*, Chester, 15 December 1923, p7; *CO*, Chester, 15 December 1923, p7, col c; R J Hickes evidence, PRO MAF 35/159

25   Stockman evidence, PRO MAF 35/165

26   Committee evidence, PRO MAF 35/162, pp104, 109

27   Stockman evidence, PRO MAF 35/165, pp92–97

28   'Disaster to Cheshire', *The Times*, London, 15 December 1923, p9, col a; 'The stock farmers' stoicism', ibid, 24 December 1923, p11, col c; 'Slaughter policy', ibid, 2 January 1924, p11, col b; 'Cattle slaughter', *Daily Mail*, London, 17 December 1923, pp8–9

29   PRO MAF 35/217

30   Stockman evidence, PRO MAF 35/165, day 2, p19

31   F Floud (1927), p129; W Mercer (1963), pp20–6

32   'Ministry faces the music', *CO*, Chester, 22 December 1923, p8; 'Resolutions versus slaughter policy carried', *CC*, Chester, 29 December 1923, p4, col f–h; ibid, p7, col a–d; ibid, p8, col a; 'Farmers condemn policy of slaughter', *The Times*, London, 21 December 1923, p10, col d; 'The stock farmers' stoicism', ibid, 24 December 1923, p11, col c

33   'Chester Cathedral: special prayers', *The Times*, London, 29 December 1923, p8, col f; 'Partial isolation in Cheshire', ibid, 16 January 1924, p9, col a; 'Farming notes', *CC*, Chester, 5 January 1924, p7

34   Dr Burton correspondence, *CO*, Chester, 15 December 1923, p12, col b; J S Thomson correspondence, *CC*, Chester, 29 December 1923, p7, col b;

'Medical views', *The Times*, London, 4 January 1924, p12, col c; 'A medical view', *CO*, Chester, 5 January 1924, p3, col c

35 Stockman evidence, PRO 35/165, p18

36 Stockman evidence, PRO 35/165, pp15–17

37 'Cattle disease', *The Times*, London, 1 January 1924, p12, col g; 'The policy of slaughter', ibid, 2 January 1924, p7, col a; 'Farmers and the Ministry', ibid, 2 January 1924, p11, col b; 'Farmers' committee meet Ministry officials', *CC*, Chester, 5 January 1924, p7; 'Cheshire farmers: open meeting with prayer', *CO*, Chester, 5 January 1924, p3

38 Cheshire County Council Diseases of Animals Sub-committee minutes, 4 January 1924, Cheshire Country Records Office

39 'Cattle Scourge', *CO*, Chester, 12 January 1924, p3, col a

40 'Slaughter policy to go on', *CC*, Chester, 12 January 1924, p7

41 Stockman evidence, PRO MAF 35/165, pp96–97

42 Holmes Chapel Farmers' Union minutes, 2 January 1924, Cheshire Country Records Office

Resolution and aftermath

43 *CO*, Chester, 12 January 1924, p7, col d

44 'FMD – slaughter policy', *The Times*, London, 17 January 1924, p8, col d

45 *CO*, Chester, 19 January 1924, p1

46 H Tollemache correspondence, *CO*, Chester, 26 January 1924, p5, col e; J Willett, 'The grim tragedy ending – what of the future?', *CC*, Chester, 26 January 1924, p7; 'FMD', *The Times*, London, 29 February 1924, p7, col c

47 NFU County Restocking Committee minutes, 20 February 1924, Rural History Centre, Reading University

48 *Annual Proceedings under the Diseases of Animals Acts* (1924)

49 Committee evidence, PRO MAF 35/168

50 Cheshire representatives' evidence, PRO MAF 35/164; Stockman evidence, PRO MAF 35/165

51 Chairman comments, PRO MAF 35/165, day 2, p9

52 *Report of the 1924 Departmental Committee on FMD* (1924)

53 Stockman evidence, PRO MAF 35/165, p50

CHAPTER 4

The rise of the international meat trade

1 Evidence of W M Furnival, F P Matthews and W Smart, Appendices and Index to *Report of the 1912 Departmental Committee on FMD* (1912); RAS Veterinary Committee minutes, July 1914–December 1915, passim, Rural History Centre, Reading University

2 The above section is drawn from P Smith (1969), ch1–5; M Machado (1969), p3; R Gravil (1970), pp147–60; R Perren (1978), ch10; P Goodwin (1981), pp29–35; D Rock (1985); D Sheinin (1994), pp501–11

Suspicions aroused?

3 G Cosco and A Aguzzi (1919)
4 W Young evidence, PRO MAF 35/166; J O Powley and J Kelland evidence, PRO MAF 35/164; *Report of the 1924 Departmental Committee on FMD* (1924), pp10, 37–38, 58–59
5 Leaflet dated 31 December 1923, PRO MAF 35/164
6 FMD Research Committee, Committee Meetings 9–15, April–December 1925 (hereafter FMDRC CM), passim; FMD Research Committee, Committee Paper 30B, March 1925 (hereafter FMDRC CP); FMDRC CM 16, February 1925; FMDRC CP 50, March 1926; Obituary: Stewart Stockman (1926)
7 *Annual Reports of Proceedings under the Diseases of Animals Acts* (1926), p24–25; FMDRC CP 54, May 1926; RAS minutes, 2 June 1926, Rural History Centre, Reading University
8 *Hansard* [HC], 9 June 1926, vol 196, col 1465–67; ibid, 17 June 1926, vol 196, col 2454–56; ibid, 23 June 1926, vol 197, col 359–62; ibid, 24 June 1916, vol 197, col 575–76; ibid, 12 July 1926, vol 198, col 1608; NFU Meat and Livestock Committee minutes, 16 June 1926, Rural History Centre, Reading University
9 RAS minutes, 2 June 1926, Rural History Centre, Reading University; London Central Markets Deputation, 3 June 1926, PRO MAF 35/206; NFU Meat and Livestock Committee minutes, 16 June 1926, Reading University Rural History Unit Archive
10 *Hansard* [HC], 17 June 1926, vol 196, col 2454–56; ibid, 28 June, 1926, vol 197, col 792–94; ibid, 5 July 1926, vol 197, col 1592–94; NFMTA Deputation, 13 July 1926, PRO MAF 35/206; 'Meat embargo', *Journal of the Ministry of Agriculture*, vol 33, 1926–1927, pp577–78
11 FMDRC CP 54, May 1926; FMDRC CM 19, June 1926; MAF correspondence with Argentine Legation, June 1926, PRO MAF 35/206
12 MAF correspondence with Robertson, November 1928, PRO MAF 35/208; P Smith (1969), ch 2 and 3; P Goodwin (1981); D Rock (1985), pp162–86; M Foran (1998); S Moore (1993)
13 D Spring (1984); J Brown (1987), ch3–5; A Cooper (1989); G Cox, P Lowe and M Winter (1991); S Moore (1991); J Brown (2000)
14 Correspondence, PRO MAF 35/206, MAF 35/210, MAF 35/211, MAF 35/121
15 FMDRC CM 23, 7 December 1926; FMDRC CP 59b, November 1926; *2nd Progress Report of the FMD Research Committee*, HMSO, London, 1927; *3rd Progress Report of the FMD Research Committee*, HMSO, London, 1928

16 *Annual Report of Proceedings under the Diseases of Animals Acts* (1927), p27
17 FMDRC CP 58, November 1926
18 FMDRC CP 59b, November 1926; FMDRC CM 23, December 1926; *Hansard* [HC], 25 November 1926, vol 200, col 527–28
19 'FMD', *The Times*, London, 9 January 1928, p20, col a; ibid, 13 February 1928, p18, col a
20 *Hansard* [HC], 28 February 1928, vol 214, col 245–46; 'Council of Agriculture for England', *Journal of the Ministry of Agriculture*, vol 34, 1927–1928, pp1127–28
21 *Hansard* [HL], 21 February 1928, vol 70, col 206; ibid, 20 March 1928, vol 70, col 513–28; ibid, 2 May 1928, col 955–67; ibid, 8 May 1928, vol 70, col 1015–28
22 P Smith (1969), pp119–21; F Capie (1981)
23 Report of Council of Agriculture meeting, January (1928), PRO MAF 35/208
24 RAS Council minutes, 2 May 1928, Rural History Centre, Reading University; *Hansard* [HL], 27 June 1928, vol 71, col 747–66
25 *Hansard* [HC], 23 April 1928, vol 216, col 613–15
26 'FMD', *The Times*, London, 6 August 1928, p16, col b; *Hansard* [HC], 1 March 1928, vol 214, col 592–94; *Hansard* [HL], 21 February 1928, vol 70, col 206; ibid, 20 March 1928, vol 70, col 513–28; ibid, 2 May 1928, col 955–67; ibid, 8 May 1928, vol 70, col 1015–28

The Argentine reaction

27 J Richelet, 'Meat inspection in the Argentine', PRO MAF 35/206; R Jackson correspondence, press notice of meeting between *frigorifico* owners, report by Robertson, PRO MAF 35/208; P Goodwin (1981), pp31, 39–42
28 Internal correspondence, January 1927, PRO MAF 35/208
29 'FMD', *The Times*, London, 8 August 1930, p7, col d; Lignieres correspondence and W H Andrews minutes, 3 February 1932, PRO MAF 35/208; M Machado (1969), pp5–10; D Spear (1982); D Sheinin (1994), pp511–20
30 See, for example, M Machado (1969).
31 Meeting, Guinness and Uriburu, 26 July 1926; Jackson to Smart, 4 August 1926; R Jackson minutes, 3 November 1927; Correspondence with Robertson, November 1928, PRO MAF 35/206; J Richelet (1929); P Smith (1969), pp1–3; C Solberg (1971), p27; D Sheinin (1994), pp518–19
32 Report of Bledisloe's visit, letter from Luis Duhau to Bledisloe, *Review of the River Plate*, 2 March 1928, PRO MAF 35/208; Bledisloe correspondence to *The Times*, London, 24 March 1928, p10, col b
33 Correspondence, January 1918, PRO MAF 35/208; German report, NFU minutes, 17 October 1928, Rural History Centre, Reading University

Twisted science

34   FMDRC CM 48, February 1930; FMDRC CM 58, April 1931
35   FMDRC CP 102C, October 1930; W Wittman (1999), p28
36   FMDRC CP 102C, October 1930; FMDRC CM 54, November 1930
37   FMDRC CP 106B, January 1931; FMDRC CM 42–56, May 1929 to February 1931, passim
38   FMDRC CP 102B, October 1930; FMDRC CP 106B, January 1931; FMDRC CM 55, December 1930
39   FMDRC CM 55, December 1930
40   H Skinner (1989)
41   FMDRC CM 73, December 1932
42   FMDRC CM 81, November 1933
43   Frood reports, PRO MAF 35/208; Memo, 'FMD: Measures to prevent introduction by carcasses from Argentina', PRO MAF 35/207
44   Correspondence, 1934–35, PRO MAF 35/207; Reports by Captain V Boyle, PRO MAF 35/209
45   Kelland minutes, 31 January 1933, List of primary FMD outbreaks 1933–34, PRO MAF 35/207; FMDRC CM 78, June 1933; FMDRC CM 81, November 1933; FMDRC 92–94, December 1934–February 1935; FMDRC CP 242, February 1935
46   Report by Cabot, October 1935, PRO MAF 35/207; Memo by Cabot, 31 December 1936, PRO MAF 35/227; Vandepeer minutes, 22 November 1937, PRO MAF 35/209; FMDRC CM 120, November 1937; FMDRC CM 122, February 1938
47   *Annual Report of Proceedings under the Diseases of Animals Acts* (1935), p55
48   T Rooth (1985), pp173–74, 189–90; F Capie (1978), pp49–51
49   Cabinet meeting, 2 November 1933, PRO CAB 27/495; NFU Meat and Livestock Committee minutes, 1933–37, passim, Rural History Centre, Reading University; P Smith (1969), pp140–49, 198–99; R Gravil (1970); P Goodwin (1981), pp43–45; D Rock (1985), pp224–31; D Sheinin (1994), p520

CHAPTER 5

The birth of scientific medicine?

1   For a classic account, see P de Kruif (1926).
2   For discussions of these issues, see S E D Shortt (1983); C Lawrence (1985); J Austoker (1988), ch 2; B Latour (1988); A Cunningham (1992); P Weindling (1992); J H Warner (1995); M Harrison (1996); J A Mendelsohn (unpublished, 1996); P Weindling (1993); S Sturdy and R Cooter (1998); C Lawrence (1998); P Baldwin (1999); L Bryder (1999); M Worboys (2000); K Waddington (2001); and J Fisher (2003)

3　H Chick, M Hume and M MacFarlane (1971); A Landsborough Thomson (1973), ch 1; G MacDonald (1980); W Foster (1983), ch 1 and 2; C Booth (1987); J Austoker (1988), ch 2; K Vernon (1990); C Booth (1993); S Sturdy and R Cooter (1998)

4　A laboratory existed at the Board of Agriculture as early as 1893; but for many years this was just a room where post-mortem specimens were examined for signs of swine fever or bovine pleuro-pneumonia. See I Pattison (1981); I Pattison (1984), ch 16; Obituary: Sir John McFadyean (1995).

5　*Report of Proceedings under the Diseases of Animals Act* (1909); Anon (1967), p62; I Pattison (1979); R Olby (1991); M Worboys (1991), pp325–26; K Vernon (1997); www.defra.gov.uk/corporate/vla/aboutus/aboutus–history.htm

Starting out: FMD research in Britain and Europe up to 1924

6　Smith Hughes (1977); A P Waterson and L Wilkinson (1978)

7　H Skinner, 'Passive immunisation against FMD, a review', FMDRC CP 411 (1940?); H Schmiedebach (1999); W Wittmann (1999)

8　Stockman evidence, *Report of the 1912 Departmental Committee on FMD* (1912)

9　Evidence, Appendices and Index to *Report of the 1912 Departmental Committee on FMD* (1912); *Report of the 1912 Departmental Committee on FMD*

10　*Report of the Departmental Committee Appointed to Inquire into FMD* (1914)

11　Correspondence, PRO MAF 35/216; 'Report on the FMD Committee', FMDRC CP 2, 1924

12　For European disease incidence, 1921–25, see *Hansard* [HC], 10 June 1926, vol 196, pp1713–14; for information on how geography affected disease control policies, see P Baldwin (1999).

13　I Pattison (1984), ch 16; H Skinner (1992); J Fisher (2003)

Doctors, vets and the purpose of scientific enquiry

14　Stockman correspondence to Dr Leishman, 7 January 1924, PRO MAF 35/217

15　See notes 2 and 3 in 'The birth of scientific medicine?'; J Eyler (1987)

16　'Editorial', *Lancet*, 8 March 1924, pp504–05

17　'Editorial', *Lancet*, 12 January 1924, pp83–85

18　'Editorial', *British Medical Journal,* 19 January 1924, pp121–22. Like MAF, the Ministry of Health believed its scientists should pursue research that served its policies. See E Higgs (2000).

19　'Medical views – divided counsels in Cheshire', *The Times*, London, 4 January 1924, p12, col c; Dr Young, Dr Peyton and Dr Grace correspondence, ibid, 7 January 1924, p18, col c

20   Peyton interview, *CO*, Chester, 5 January 1924, p3, col c
21   Ibid; Dr W Hodgson, evidence to 1924 committee of inquiry, PRO MAF 35/164; Obituary: Professor Beattie (1955)
22   J M Beattie and D Peden, 'FMD in rats', *Lancet*, 2 February 1924, pp221–22
23   Stockman evidence to 1924 committee of inquiry, PRO MAF 35/165
24   Obituary: Sir W M Fletcher, (1932–35)
25   Fletcher correspondence to Lord Mildmay, 1 April 1924, PRO FD 1/1346
26   Evidence, Appendices, Index and *Report of the Departmental Committee Appointed to Inquire into the Requirements of the Public Services with Regard to Officers Possessing Veterinary Qualifications* (1912–1913); Development Commission Committee, 1913, PRO D4/91; 'Swine fever – need for research', *The Times*, London, 4 May 1914, p4, col a; Sir C Allbutt correspondence to *Veterinary Record*, 26 June 1920, pp609–11; ongoing correspondence, ibid, 1920–21, passim; Development Commission Advisory Committee, 1920–22, PRO FD 1/4364; MRC Committee on RVC Charter, 1922, PRO FD 1/5048; Stockman correspondence to Dr Leishman, 7 January 1924, PRO MAF 35/217; Sir C Allbutt, 'Inaugural address', *Veterinary Record*, 12 January 1924, pp17–18; RD letter to *The Times*, London, 8 January 1924, p8, col c; H A Reid letter, ibid, 14 January 1924, p14, col b; I Pattison (1979); I Pattison (1984); R Olby (1991); T DeJager (1993); K Vernon (1997); M Worboys (1991)
27   J Austoker (1988), ch 3; T DeJager (1993); E Higgs (2000)
28   MRC Committee on RVC Charter, 1922, PRO FD 1/5048; Fletcher correspondence, January 1922, PRO FD 1/4364
29   A Kraft (unpublished, 2003)
30   G Adami, correspondence to *The Times*, quoted in *Veterinary Record*, vol 33, 1920–1921, pp21–22
31   'Sir Stewart Stockman on the FMD problem' *Veterinary Record*, 19 January 1924, pp35–39
32   'Editorial', *Veterinary Journal*, February 1924, pp56–57
33   'Editorial', *Veterinary Record*, 19 January 1924, p49
34   'Sir Stewart Stockman on the FMD problem', *Veterinary Record*, 19 January 1924, p35
35   Cabinet Committee meetings, PRO MAF 35/217; Fletcher letter to C Sherrington, 22 February 1924, PRO FD 1/1346
36   Fletcher correspondence with Leishman, August 1921, PRO FD 1/4364; MRC Committee on RVC Charter, PRO FD 1/5048
37   Fletcher letter to C Sherrington, 22 February 1924, PRO FD 1/1346
38   Stockman letter to Lord Ernle, 7 January 1924, PRO MAF 35/217
39   Stockman letter to Leishman, 7 January 1924; Leishman report, PRO MAF 35/217
40   Stockman response, 8 February 1924, PRO MAF 35/217
41   Fletcher letters to C Sherrington, 22 and 25 February 1924, PRO FD 1/1346
42   'FMD', *The Times*, London, 29 February 1924, p7, col c; Obituary: William Leishman (1995)

43   Correspondence, PRO MAF 33/64; K Angus (1990), ch 1 and 2; E Tansey (1994)

Keeping FMD out, 1924–1938

44   A Woods interview with H H Skinner, 7 March 2000; H Skinner (198–, unpublished), section 1
45   ibid; FMDRC CM 1924–37, passim
46   FMDRC CP 4, 29 February 1929; FMDRC CM 19, 30 March 1926; FMDRC CP 20, 6 July 1926
47   Correspondence, PRO MAF 33/533; T DeJager (1993), p147
48   'FMD' *The Times*, London, 29 February 1924, p7, col c
49   For a chronology of scientific breakthroughs, see F Brown (2003).
50   ibid; FMDRC CM 1924–25, passim
51   FMDRC CM 1925–39, passim
52   *Annual Report of Proceedings under the Diseases of Animals Acts,* (1925), p17; ibid, 1928, p17; ibid, 1932, pp34–35
53   'Foot and mouth disease', *The Times*, London, 25 September 1930, p9, col c; 'Serum treatment in connection with FMD', *Veterinary Record,* 6 December 1930
54   Kelland minutes to H Dale, December 1932, PRO MAF 35/226
55   FMDRC CM 1925–33, passim; *Annual Report of Proceedings under the Diseases of Animals Acts* (1930), p8; ibid, 1931, p26; ibid, 1932, p11; ibid, 1933, p24; Andrews memo to Kelland, 30 June 1931, PRO MAF 35/266; FMDRC CP 111B, August 1931
56   FMDRC correspondence, PRO MAF 35/226
57   FMDRC CM and FMDRC CP, 1925–37, passim; Report of the ARC Committee appointed to review the work of the FMDRC, July 1933, PRO MAF 33/554
58   Report of the sub-committee of the ARC appointed to review the research work in progress on FMD, July 1937, PRO MAF 33/482
59   FMDRC CP 360, October 1938; Edwards married McFadyean's second daughter in 1922 and, hence, was Stockman's brother-in-law. The marriage did not survive. Memoir, J T G Edwards (1953)
60   *1st–5th Progress Reports of the FMD Research Committee* (1925–37); 'A note on the reporting of findings in the early years of FMD research' in H Skinner (198–, unpublished); A Woods, interview with H Skinner, 7 March 2000

A wartime threat

61   Report of Sub-committee on Biological Warfare, March 1937, PRO MAF 35/231
62   Cabot memo, April 1937, note by Market Division, May 1937, PRO MAF 35/231

63　FMDRC CP 351, 2 June 1938
64　Cabot paper, 13 November 1939, PRO MAF 250/126
65　I Galloway, 'Memo on reports of a new method of active immunisation', FMDRC CP 355, 30 June 1938; FMDRC CM 134, 31 October 1939; Cabot correspondence, May 1939, PRO MAF 35/231; Memo, 'FMD: preparation of serum', PRO MAF 250/126
66　Reports resulting from Skinner's visit, box file 8, IAH Pirbright archive; FMDRC CM 133, 27 July 1939; FMDRC CM 134, 31 October 1939; FMDRC CM 139, 7 August 1940; FMDRC CM 142, 20 May 1941; FMDRC CP 398, 1939; Rinderpest research at Pirbright, 1940–42, box file 9, IAH Pirbright archive; Memo, 'FMD: preparation of serum', PRO MAF 250/126
67　FMDRC CM and FMDRC CP, 1938–42, passim; Progress report, 1944–45, box file 9, IAH Pirbright archive
68　J Francis letter to H Skinner, 24 July 1942, box file 8, IAH Pirbright archive; 'A Note on the reporting of findings in the early years of FMD research' in H Skinner (198–, unpublished); H Skinner (198–, unpublished), section 1. Brooksby later became director of Pirbright, while Henderson went on to direct the Pan-American FMD Centre in Rio de Janeiro, Brazil, and became secretary of the ARC in 1972. A Woods interview with H Skinner, 7 March 2000
69　B Bernstein (1987); B Balmer and G Carter (1999), pp309–10
70　W A Stewart, 'Pig keeping in wartime', *Journal of the Ministry of Agriculture*, vol 46, 1939–1940, p627; 'Pig feeding with swill', ibid, pp692–93; 'Pigs on every farm', ibid, pp170–71; *Report of Proceedings under the Diseases of Animals Acts* (1938–1947), pp4–5
71　'Foot and mouth disease', *The Times*, London, 6 March 1941, p5, col e; 'Foot and mouth disease', ibid, 7 March 1941, p5, col e; *Annual Report of Proceedings under the Diseases of Animals Acts* (1938–1947), p2
72　Memos and correspondence, PRO MAF 88/295, MAF 88/296 and MAF 35/696
73　PRO WO 208/3973
74　'Nazis planned to use virus against Britain', *The Times*, London, 12 March 2001, p6, col f–g

The Cold War and biological weapons research

75　M Dando (1994), p80; G Carter and G Pearson (1996); B Balmer (1997), pp117–18; B Balmer and G Carter (1999), pp311–12
76　Research on FMD, DRP/P (51) 62, 23 August 1951, PRO DEFE 10/30; Clandestine attack on crops and livestock of the commonwealth, DRP/P (51) 77, 5 October 1951, PRO DEFE 10/30; 'Canada: FMD', *The Times*, London, 1 March 1952, p5, col c; PRO WO 195/12458; K Harrison-Jones (MAF) to Home Office, 26 February 1955, PRO MAF 250/163
77　DRP/P (51) 62, 23 August 1951, PRO DEFE 10/30
78　ibid; Bartlett to Harrison-Jones (MAF), 3 April 1952, PRO MAF 250/163

79 PRO WO 195/11426; DRP/P (51) 62, 23 August 1951, PRO DEFE 10/30; B Balmer (1997), p121, states that in the absence of accurate intelligence, much biological weapons research was shaped by the rationale that if British scientists could achieve something, then the enemy probably could, as well. This represented a self-perpetuating justification for offensive research.

80 PRO WO 195/10780

81 Wilcox minutes to Manktelow, 12 December 1951, PRO MAF 250/163

82 Correspondence, PRO MAF 117/394

83 Sir H Parker, 6 December 1951, PRO MAF 250/163

84 Sir D Vandepeer, 7 January 1952, PRO MAF 250/163

85 Correspondence, December 1951–May 1952, PRO MAF 250/163

86 Correspondence, PRO MAF 250/163, MAF 117/203, MAF 117/222, MAF 117/391 and MAF 117/394.

87 Gowers correspondence to Nugent, 15 July 1954, PRO MAF 117/394

88 DRP/P (54), 8 Oct 1954, PRO DEFE 10/33; B Balmer (1997), pp119–33; B Balmer and G Carter (1999), pp310–11

89 PRO WO 219/1421, WO 195/14245

90 *Report of the ARC 1956–1957* (1956–1957), pp101–04; ibid, 1957–1958, p133

91 The FAO was established in 1945 with a mandate to raise levels of nutrition and standards of living, improve agricultural productivity and better the condition of rural populations. It was initially headed by Sir John Boyd Orr, who tried to coordinate international action against the looming world food shortage, while also considering long-term problems in food production, distribution and consumption. J H Locke, 'FAO – an experiment in international cooperation', *Journal of the Ministry of Agriculture,* vol 53, 1946–1947, pp381–85; FAO website, www.fao.org/UNFAO/e/wmain-e.htm

92 Correspondence, PRO MAF 35/868, MAF 35/869 and MAF 252/48; *Constitution of the European Commission for the Control of Foot-and-Mouth Disease* (1953)

93 PRO MAF 252/165

94 Anon (1978), p193

CHAPTER 6

FMD returns

1 'Lord Iveagh, herd of Guernseys destroyed', *The Times,* London, 11 August 1945, p2, col d; 'Lord Iveagh, pedigree Guernsey herd slaughtered', ibid, p6, col e; Lord Iveagh correspondence, ibid, 3 July 1952, p7, col e

2 *Animal Health Services Report* (1952); W Wilson and R C Matheson (1952–1953)

3 *Report of the 1922 Departmental Committee on FMD* (1922); *Report of the 1924 Departmental Committee on FMD* (1924); PRO MAF 35/153 and MAF 35/170

4   P Self and H J Storing (1963); J Martin (2000), pp67–83; J Brown (2000)

5   *Animal Health Services Report* (1952); W Wilson and R C Matheson (1952–1953)

6   M Tracy (1989), ch 11; J Bowers (1985), p66; The FMD epidemic coincided with a balance-of-payments crisis, which the Conservative government tried to manage by repeated increases in the bank rate. The crisis eventually resolved itself as import prices fell; C Schenk (1998)

7   'Foot and mouth policy under fire', *Daily Telegraph*, London, June 1952, IAH Pirbright Archive

8   Lord Bledisloe correspondence to *The Times*, London, 7 May 1952, p7, col e; 'FMD: Kent farmers' protest', ibid, 4 July 1952, p6, col a

9   'FMD' *Daily Telegraph*, London, May 1952, IAH Pirbright Archive

10   W Rees (undated) unpublished, pp1–2

11   Frenkel discovered that virus could be cultured in-vitro using tongue tissue. This meant that scientists no longer relied upon the slow, expensive method of harvesting virus from the tongues of live infected cattle. T Dalling, evidence to Gowers committee, PRO MAF 387/18; Anon (1978), p194

12   Gower committee meetings, PRO MAF 387/28; *Report of the Departmental Committee on FMD, 1952–4*, pp135–9

13   'FMD in France: outbreak spreading', *The Times*, London, 14 July 1952, p6 col f; W Rees (date unknown, unpublished), pp1–2

14   Evidence to Gowers committee, PRO MAF 387/28 and MAF 387/31; Anon (1978), p194

15   Examples include J T Davies correspondence to *The Times*, London, 5 May 1952, p7, col g; 'Report from Denmark', *Daily Telegraph,* London, 30 June 1952, IAH Pirbright Archive; W D Thomas, 'Danish Cattle Policy', ibid, June 1952; Lord Iveagh correspondence to *The Times*, London, 3 July 1952, p7, col e; 'FMD: France', ibid, 5 September 1952, p5, col a; Chapman Pincher, 'FMD', *Daily Express*, London, 13 November 1952, PRO CAB 124/1562

Calls to vaccinate

16   Lord Bledisloe correspondence to *The Times,* London, 7 May 1952, p7, col e; Lord Iveagh evidence to Gowers committee, PRO MAF 387/7; Gloucestershire Cattle Society evidence to Gowers committee, PRO MAF 387/16

17   *Sunday Express*, London, 15 June 1952, IAH Pirbright Archive

18   George Villiers correspondence to *Daily Telegraph*, London 19 April 1952; 'Is there no better answer?', *Daily Express*, London, 15 June 1952; 'Report from Denmark', *Daily Telegraph,* London, 30 June 1952, all from IAH Pirbright Archive; Lord Iveagh correspondence to *The Times,* London, 3 July 1952, p7, col e; 'Let's look again', *Daily Mail*, London, 12 July 1952, PRO MAF 255/38; 'Vaccination: correspondence', *The Times*, London, 11 August 1952, p9, col b

19   W M Crofton (1936)

20 W M Crofton article in the *Spectator*, December 1938, PRO FD 1/1346; W Crofton correspondence to *The Times*, London, 7 March 1941, p5, col e; Crofton practised vaccine therapy, a controversial and largely discredited therapeutic measure. See M Worboys (1992)

21 Correspondence, PRO MAF 35/452 and 453

22 W Crofton correspondence to *Daily Telegraph*, London, March 1952, IAH Pirbright Archive

23 'FMD', *Daily Telegraph*, London, May 1952, IAH Pirbright Archive

24 'Foot and mouth policy', *Daily Telegraph*, London, 5 June 1952; 'Foot and mouth policy under fire', ibid, 12 June 1952; 'Is there no better answer?', *Sunday Express*, London, 15 June 1952; 'Report from Denmark', *Daily Telegraph*, London, 30 June 1952, all from IAH Pirbright Archive; Lord Iveagh correspondence to *The Times*, London, 3 July 1952, p7, col e; 'Let's look again', *Daily Mail*, London, 12 July 1952, PRO MAF 255/174; 'FMD', *Veterinary Record*, 16 August 1952, p481

The ministry stands firm

25 'Work with Mexican field strains, 1946–47', IAH Pirbirght Archive; I Galloway, 'Draft Report, 1937–1952', IAH Pirbright Archive, pp19–22, 30–31

26 This transfer of control resulted from a dispute between MAF officials and the FMDRC, which was reconstituted after the war under the chairman, Sir Alan Drury. The FMDRC felt that its pre-war freedom of action had been curtailed, and when, in 1948, the Treasury intervened to prevent it from raising scientists' salaries, many members threatened to resign. Fearing adverse publicity, MAF agreed to transfer control of Pirbright to the ARC, and to appoint the existing FMDRC as its governing body, a move that brought Pirbright into line with the other state-aided agricultural research institutes. MAF retained control over financing and continued to station an officer at Pirbright; but the ARC became responsible for the research programme and for staffing. One condition of the change was the continued appointment of the CVO to the governing body. See PRO MAF 33/944 and MAF 117/73.

27 NFU statement, *The Times*, London, 10 December 1951, p2, col e; W R Wooldridge, 'FMD: slaughter policy justified', ibid, 17 December 1951, p3, col c; J Salter-Chalker correspondence to *Daily Telegraph*, London, March 1952, IAH Pirbright Archive; Memo, 'FMD – appreciation of present position', 8 May 1952, PRO MAF 255/174; Sir Merrik Burrell correspondence to *The Times*, London, 12 July 1952, p5, col f; J Turner, 'Why slaughter must go on', *News of the World*, London, 11 May 1952, PRO MAF 255/174; 'FMD', *Veterinary Record*, 16 August 1952, p481; Lord Rothschild, 'FMD', *Sunday Times*, London, 5 October 1952. Rothschild wrote this article after discussing the matter with MAF officials; PRO MAF 124/1562

28 PRO MAF 250/163

29　W Thomas, 'Fighting FMD', *Daily Telegraph*, London, 25 March 1952, IAH Pirbright Archive; 'FMD', *The Times,* 28 March 1952, p7, col d; ibid, 9 May 1952, p7, col c; '"Forceps": foot and mouth disease and its control', *Sport and Country,* 16 April 1952, IAH Pirbright Archive; 'Minister's statement on slaughter policy', *The Times,* London, 3 May 1952, p3, col b; T Dugdale, Parliamentary speech, ibid, 9 May 1952, p6, col g; T Dugdale answers Parliamentary Questions, *Veterinary Record,* 17 May 1952, p296; 'FMD research' and 'Ban on cattle from channel islands', *Daily Telegraph,* June 1952, IAH Pirbright Archive; I Galloway, 'FMD', *British Agricultural Bulletin,* July 1952, p67; 'ARC report', *The Times,* London, 28 August 1952, p2, col e

30　'FMD policy – comparisons with the Continent', *The Times,* London, 12 May 1952, p3, col d; 'The present FMD position: BVA statement', *Veterinary Record,* 17 May 1952, p295; 'Vaccination fails to save European cattle', *Sunday Times,* London, 29 June 1952, IAH Pirbright Archive; T Dugdale in Parliament, *Veterinary Record,* 5 July 1952, p403; *FMD Research – Interim Report* (1952), pp1–2

31　Evidence drawn from FMDRC CM, 1938–1941, passim; Lab 2 correspondence, 1940–1942, box file 8, IAH Pirbright Archive; 'Progress report, 1944–1945', box file 9, IAH Pirbright Archive; FMDRC meeting, 25 March 1946, IAH Pirbright Archive; I Galloway, 'Considerations of some important aspects of recent investigations on FMD', cited in H Skinner (198–, unpublished), section 10; I Galloway, 'Draft report for 1935–1952', box file 9, IAH Pirbright Archive, pp14–19; W Henderson (1985), pp10–22

32　Metaphors drawn from *The Times* and *Daily Telegraph*, London, March–August 1952, passim; for striking examples of military metaphors employed during the 1922–1924 epidemics, see comments by Stockman, German and Pretyman, PRO MAF 35/164 and MAF 35/165; for their use during the 2001 epidemic, see B Nerlich, C Hamilton, and V Rowe (2001)

33　W Wooldridge correspondence to *The Times*, London, 9 May 1952, p 7, col e; Evidence of J Ritchie and W Tame to Gowers committee, PRO MAF 387/2; Gowers committee visit to France and Switzerland, PRO MAF 387/23; Evidence of T Dalling to Gowers committee, PRO MAF 387/18; Gowers committee meetings, PRO MAF 387/28; Dr Simms correspondence, PRO MAF 387/31

34　Gowers committee meetings, PRO MAF 387/23

The dispute continues

35　H Skinner (1993)

36　E Ratcliffe, correspondence to *Daily Telegraph,* London, March 1952, IAH Pirbright Archive; 'Foot and mouth', *Economist,* London, 24 May 1952, p506; 'Foot and mouth policy under fire', *Daily Telegraph,* London, June 1952, IAH Pirbright Archive; 'Report from Denmark', ibid, 30 June 1952; Lord Iveagh correspondence to *The Times,* London, 3 July 1952, p7, col e; A M Allen,

'Immunised cattle', *Daily Telegraph*, London, October 1952, IAH Pirbright Archive; George Villiers, 'The case for inoculation as an aid to the fight against FMD', PRO MAF 35/866

37 'FMD', *Daily Telegraph*, London, 5 June 1952, IAH Pirbright archive

38 *FMD Research – Interim Report* (1952). The original foreword to this report, drafted by Lord Woolton, stated that the report was intended 'to provide assurance that the work is being prosecuted both vigorously and on lines likely to lead to results of practical value.' The last phrase was removed at the ARC's suggestion. PRO CAB 124/1562

39 'FMD', *Daily Telegraph*, London 28 August 1952, IAH Pirbright Archive

40 Gowers committee meetings, PRO MAF 387/23

41 'Foot and mouth policy under fire', *Daily Telegraph*, London, 12 June 1952, IAH Pirbright Archive

42 George Villiers, 'The case for inoculation as an aid to the fight against FMD', PRO MAF 35/866

43 Evidence of R H Bathurst to Gowers committee, PRO MAF 387/16

44 Galloway correspondence to MAF, 26 June 1952, PRO MAF 35/866

The Gowers Committee of Inquiry

45 'FMD', *The Times*, London, 2 August 1952, p7, col d

46 PRO MAF 387/1

47 Gowers committee meeting, PRO MAF 387/2

48 PRO MAF 387/1

49 *Report of the Departmental Committee on FMD, 1952–1954* (1952–1954), pp55–56; PRO MAF 387/7 and MAF 387/9; Galloway apparently prevented further tests from going ahead.

50 *Report of the Departmental Committee on FMD, 1952–1954* (1952–1954), pp6–7

51 *Report of the Departmental Committee on FMD, 1952–1954* (1952–1954), pp40–41, appendix xv

52 *Report of the Departmental Committee on FMD, 1952–1954* (1952–1954), p41

53 *Report of the Departmental Committee on FMD, 1952–1954* (1952–1954), pp21–35, 41–57, 135–39

54 PRO MAF 387/24

55 Gowers committee meetings, PRO MAF 387/28

56 PRO MAF 387/14 and MAF 387/18; *Report of the Departmental Committee on FMD, 1952–1954* (1952–1954), p35; M Machado (1968); L Lomnitz and L Mayer (1994)

57 ibid, pp21–35; Gowers committee meetings, PRO MAF 387/18, MAF 387/28 and MAF 387/24

58 *Report of the Departmental Committee on FMD, 1952–1954* (1952–1954), p50

59 *Report of the Departmental Committee on FMD, 1952–1954* (1952–1954), pp52–54; Gower to Thorne, 8 March 1954, PRO MAF 387/35

60   Tame minutes, 17 March 1955, PRO MAF 35/872
61   Dunnett minutes, 19 March 1955, PRO MAF 35/872
62   MAFF press release, *The Times*, London, 29 July 1954, p4, col c
63   'The Departmental Committee on Foot-and-Mouth Disease', *Veterinary Record*, 30 April 1955, pp327–28

CHAPTER 7

Setting the scene

1   A phrase coined by Prime Minister Harold MacMillan during the later 1950s.
2   Anon (1965); E Madden (1984); J Bowers (1985); D Grigg (1987), pp183–87; P Brassley (2000b); J Martin (2000). To learn how the NFU managed to gain such power over agricultural policy, see P Self and H Storing (1963), G Cox, P Lowe and M Winter (1986); M Smith (1990).
3   H W Steele-Bodger, 'Presidential Address', *Veterinary Record*, 16 November 1940, pp795–98; *Second Report of the Committee on Veterinary Education in Great Britain* (1944), pp3–6; Memoir, Henry William Steele-Bodger (1952); W Wooldridge (1954); Anon (1965), pp214–35; Obituary: Dr W R Wooldridge (1966); E Madden (1984); I Pattison (1984), ch 19 and 20; Interviews with Mary Brancker, FRCVS, 4 November 2002, and Professor Alisdair Steele-Bodger, FRCVS, 27 January 2003. The old system whereby veterinary inspectors were attached either to MAF or to the local authorities ended in 1938, when a unified, centralized veterinary administration was established. See A Hardy (2003).
4   'FMD', *The Times*, London, 28 October 1967, p3, col d
5   *Report of the Committee of Inquiry into FMD* (1968–1969), p867

The view from the ground

6   *Animal Health Services Report* (1967), p10
7   'Funeral pyres light lifeless landscape', *Daily Telegraph*, London, 7 November 1967, p17, col a
8   G Brooke correspondence to *Daily Telegraph*, London, 24 November 1967, p13, col a
9   The epidemic dealt a second blow to rural pubs. Business had already dropped off when the police force adopted the breathalyser to detect drunken motorists.
10 The above three paragraphs are drawn from the *Daily Telegraph*, London, and the *Chester Chronicle*, Chester, November–December 1967, passim; H Hughes and J Jones (1969); R Whitlock (1969); *Animal Health Services Report* (1968)
11 'Controls over outbreaks: statutory procedures', PRO MAF 287/497; Note by T Ivey, 8 February 1968, PRO MAF 287/482

12 'What F&M means', *Chester Chronicle*, Chester, November 24, p9; Correspondence, PRO MAF 287/512

13 Correspondence with MOD, November 1967, PRO MAF 394/11

14 Cheshire and Devon County Council correspondence, November 1967–January 1968, PRO MAF 287/474; Cheshire County Council, Memo of Evidence to the Committee, May 1968, PRO MAF 287/518; NFU Press release, 9 August 1968, NFU Evidence to Northumberland Inquiry, PRO MAF 287/519

15 *1st draft: Interim Report of Working Party*, 1968, PRO MAF 287/476; Memo of internal working party on FMD, 29 May 1968, PRO MAF 283/693

16 R Hinks memo, PRO MAF 287/484

17 'Provision of information and advice', PRO MAF 287/484; Memo, 'FMD epidemic, arrangements at Ministry HQ', PRO MAF 287/511

18 The above two paragraphs are taken from *Daily Telegraph*, London, and *Chester Chronicle*, Chester, November to December 1967, passim; W Jones correspondence to Jasper More, 8 December 1967, PRO MAF 287/488; 'Foot and mouth disease', *The Times,* London, 23 April 1968, p4, col g; *Report of the Committee of Inquiry into FMD*, 1968–1969, pp15–27, 56

19 Correspondence and memos, PRO MAF 287/484

20 *Hansard* [HC], 27 November 1967, vol 756, col 34–42; Peak District National Park's Statement of Evidence, May 1968, PRO MAF 287/518; NFU evidence to the Northumberland Inquiry, June 1968, PRO MAF 287/519. Ireland had been free of FMD since 1942, and was keen to remain that way as 60 per cent of its export earnings came from the livestock trade. Blaney went on to launch a 'Stay in Britain for Christmas' campaign to persuade Irish migrant workers against returning home. PRO FCO 23/186

21 Parliamentary questions, *Hansard* [HC], 28 November–8 December, passim; Agriculture debate, ibid, 4 December, vol 755, col 969–1086; *Chester Chronicle, Daily Telegraph* and *Daily Mail* coverage, 28 November–11 December, passim; 'Restrictions on the community', PRO MAF 287/486; FMD 'Temporary Restrictions' Order, 8 December 1967, J Hensley (MAFF) letter to Knowles (NFU), 21 December 1967, PRO MAF 287/488; 'Emergency powers – submission to minister', PRO MAF 394/11

22 'Inquiry into farm epidemic', *The Times,* London, 5 December 1967, p6, col d

23 'End of the Cheshire we knew?' *Chester Chronicle*, Chester, 15 December 1967, p1

24 'It takes some nerve to stand this', *Chester Chronicle,* Chester, 1 December 1967, p1

25 Ernest Dewhurst, 'Farms' lesson from the past', *Guardian,* London, 6 December 1967, p5

26 R F Laurence, 'Buried in the country', *Guardian,* London, 23 December 1967, p5

27 *Chester Chronicle*, December 1967, passim; 'Farms' lesson from the past', *Guardian*, London, 6 December 1967, p5; 'F&M farms get crop grant', *Daily Telegraph*, London, 11 December 1967, p1; 'Valuation and compensation', PRO MAF 287/499

28   'A savage memo concerning the Peak District', PRO MAF 287/488; 'Tourists' Mecca now a ghost town', *Daily Telegraph*, 2 February 1968, p23

29   E H Bott, 'Huntin', Shootin' n' Fishin'', 4 January 1968, PRO MAF, 287/486

30   J Hensley to W Tame, 5 January 1968, PRO MAF 287/486; E H Bott to W Tame, January–February 1968, PRO MAF 287/487; Correspondence, PRO MAF 287/488. The Northumberland Committee of Inquiry into the epidemic eventually decided that controls should be based upon veterinary, not political, considerations 'and produce as little disturbance of normal commercial and public activities as possible'. Sadly, such considerations were not taken into account in 2001. *Report of the Committee of Inquiry into FMD* (1969–1970), p79

31   *Chester Chronicle*, Chester, February–March 1968, passim

32   Other members included NFU Vice-President Lord Plumb; William Weipers, head of veterinary education at Glasgow University; Mr Cripps, recorder of Nottingham; Professor D Evans, FRS, professor of bacteriology and immunology at London School of Hygiene and Tropical Medicine; Professor D Walker, professor of economics at the University of Exeter; and Sir Edward Thomson, chair of Allied Breweries. See PRO MAF 287/522.

The vaccination question

33   Examples include the Rural Dean of Bickley's letter to the *Chester Chronicle*, Chester, 8 December 1967, p9; Edward Coleman correspondence to *Daily Telegraph*, London, 14 November 1967, p16; G Brooke correspondence, ibid, 24 November 1967, p13, col a; E J Richmond correspondence to *Economist*, London, 25 November 1967, p812. See also 'Vaccine conflict as farm toll rises', *Daily Telegraph*, London, 27 November, p7; 'End farm slaughter says RSPCA', ibid, 28 November 1967, p15; 'Editorial', ibid, 29 November 1967, p16; 'Life on the farm in Cheshire', *Chester Chronicle*, Chester, 8 December 1967, p9; Walter Hume, 'It's time to change the policy', ibid, 5 January 1968, p25

34   Quotations drawn from *Hansard* [HC], *Daily Telegraph*, *Daily Mail*, *Chester Chronicle*, 20 November–23 December 1967, passim

35   'Emergency: this runaway plague is putting our food in peril', *Daily Mail*, London, 24 November 1967, p6; Northumberland Committee of Inquiry minutes, 5 June 1968, PRO MAF 287/492

36   D Christie, L Reynolds and E Tansey (2003), p52

37   'Farm plague battle is stepped up', *Daily Mail*, London, 27 November 1967, p1; Peart reply to Parliamentary Question, *Hansard* [HC], 28 November 1967, vol 755, col 70–71

38   Minister's Broadcast, PRO MAF 394/11

39   'Report from Parliament', *Guardian*, London, 5 December 1967, p2; D H Smith correspondence, 1 December 1967, PRO MAF 287/461

40   Mary Brancker was the first woman to hold this post. For more details of her life and work, see M Brancker (1972).

41 D Christie, L Reynolds and E Tansey (2003), p9
42 'Ministry has no plans to vaccinate', *Guardian*, London, 3 December 1967, p14; 'Memo: To RVOs, DVOs and DEOs concerned', PRO MAF 287/461; FMD Area Vaccination Scheme, December 1967, PRO MAF 287/479/1
43 G Amos, 'Personal Views for oral hearing of the Northumberland committee', 9 April 1968, PRO MAF 287/505/1
44 Correspondence, PRO MAF 287/461, MAF 287/462, MAF 287/479/1; E H Bott minutes, 17 July 1968, Reid minutes, 24 January 1969, PRO MAF 287/479/1; Tame minutes, 26 September 1968, PRO MAF 287/503
45 Carnochan minutes, 22 July 1968, PRO MAF 276/476; Northumberland Committee of Inquiry minutes, 5 June 1968, PRO MAF 287/92; Committee of Inquiry minutes and 'Paper VII: Vaccination Policies', PRO MAF 287/501; Correspondence, September 1968, MAF 287/503

To import or not to import? The meat question

46 Correspondence, PRO MAF 88/295, 88/296, 35/696; 'Foot and mouth disease, Argentina: NFU discussion', *The Times,* London, 23 January 1958, p5, col e; ibid, 17 February 1958, p14, col c; 'Foot and mouth disease, Argentina: discussed in Parliament', ibid, 6 March 1958, p14, col g
47 'Foot and mouth disease – preventive measures', *The Times,* London, 28 January 1957, p4, col 5. Peron set out to restore Argentina to national and international greatness. He believed that developed nations had a vested interest in preventing Argentine industrialization because they wanted to continue exporting manufacturing goods in return for Argentine agricultural produce. He tried to challenge this system by turning his country into an independent industrial power. His interventions diverted capital out of agriculture and into manufacturing and brought to an end the political and economic hegemony of the rural aristocracy. Although later regimes attempted to encourage agriculture, cattle ranchers remained suspicious of their motives and the industry failed to expand. G Wynia (1978), p240–41; D Rock (1985), p298
48 MAF, C Jewell, T Dalling and P Ellis, Evidence to the Gowers committee, PRO MAF 387/2, 387/4, 387/18 and 387/25; R P Burgess, correspondence to *The Times,* London, 25 November 1957, p2, col f; ibid, 24 March 1958, p2, col f; C Jewell, correspondence to the *Daily Telegraph*, 14 February 1958, PRO MAF 35/696; J Ritchie, 'FMD in South America,' *Journal of the Ministry of Agriculture,* vol 66, 1959–1960, pp324–27
49 Correspondence to John Hare, 12 February 1958, PRO MAF 35/696; Letter from Anthony Hurd, 18 March 1958 and Report of John Ritchie's trip to South America, 4 June 1959, PRO MAF 255/891
50 R P Burgess, letter to *The Times,* London, 25 November 1957, p2, col f; *Farmer and Stockbreeder,* 3 October 1957; For more information see L Randall (1978), pp231–34; E Milenky (1978), pp11–19; C Watson (1984), p64–65; D Cavallo and U Mundlak (1982), p20; D Rock (1985), ch 8

51   'Foot and mouth disease – Parliament questions,' *The Times*, London, 1 November 1957, p4, col f; AHD memo on Argentine FMD, 9 December 1957, PRO MAF 35/696; Parliamentary Questions, 26 February 1958 and 5 March 1958, PRO CAB 124/1563; Farmers' reply, 5 February 1958, PRO MAF 35/696. For details of a failed 1958 attempt to reduce Argentine meat imports, see A Woods (2002), pp243–44.

52   *Animal Health Services Report* (1960). The pork trade was relatively expendable because British pig production was booming. It also presented greater dangers than the beef trade because pigs were not vaccinated against FMD and their movements were not subject to the Bledisloe agreement. 'Livestock: foot and mouth disease: discussed', *The Times,* London, 17 November 1960, p7, col a; J Martin (2000) pp83, 87.

53   Hurd to Hare, 2 February 1960, PRO MAF 255/891; Note for Minister, 15 November 1960, and 'News from Argentina,' PRO MAF 255/891; *Animal Health Services Report* (1961–1965), passim. Another source suggests that the Argentine government's actions were motivated also by the US government's 1959 ban upon wet salt-cured meat from FMD-infected countries. See A Beynon, *The FMD situation in Argentina* (1968), PRO MAF 287/520/1

54   'The case against South America,' *Farmers Weekly*, 10 November 1967, PRO MAF 246/282; 'F&M may cost agriculture more the £12m', *Guardian* 16 November, p5; National Cattle Breeders Association letter, 20 November 1967, PRO MAF 287/474; 'Meat imports', *Daily Telegraph*, 5 December 1967, p14; General correspondence, November 1967, PRO MAF 246/282

55   See *Hansard* [HC], 14 November, vol 754, col 221–22; ibid, 20 November, col 929–31; ibid, 27 November, vol 755, col 38–42; ibid 28 November, col 70; ibid, 29 November, col 94–98

56   Diggines to Beith, 23 November 1967, Cresswell to Foreign Office, 24 November 1967, PRO FCO 7/1070

57   The sovereignty of these islands was a long-disputed matter. In 1966, the British government secretly indicated its willingness to discuss handing them over to Argentina. At the time that the meat ban was imposed, Foreign Secretary Michael Stewart was attempting to produce a politically acceptable formula for this handover. His plan was eventually frustrated by the islanders, who, on learning of the British government's intentions in March 1968, declared their furious opposition to becoming part of Argentina and lobbied the British press and Parliament for support. See G Dillon (1989), pp1–3.

58   Memo: Cabinet ministerial committee, meeting 29 November 1967, PRO FCO 7/1070

59   Agriculture Debate, *Hansard* [HC], 4 December 1967, vol 755, col 998–99; Sir Burke Trend to PM, 29 November 1967, PRO PREM 13/1934; Correspondence, PRO MAF 287/511

60   'Meat imports', *Daily Telegraph*, 5 December 1967, p14

61   'Argentina stop bids at our sales', *Daily Mail*, 6 December 1967, p11

62   Correspondence, December 1967–February 1968, PRO FCO 7/168 and FCO 7/1070

63 'Further measures to help exports', *Guardian,* 14 December 1967, p1; *Economist,* 16 December 1967, p1117; J Thompson (1970)

64 Cresswell to FO, 20 November 1967, PRO FCO 7/1070; Correspondence and press cuttings, December 1967–February 1968, PRO FCO 7/168

65 For evidence of the leeway granted to Argentine establishments that was found to have contravened public health regulations, see PRO MAF 276/306 and 276/261. This laxity led to the importation of contaminated tins of corned beef, which caused a 1964 outbreak of typhoid in Aberdeen. Four hundred people contracted the disease. *The Aberdeen Typhoid Outbreak of 1964: Report of the Departmental Committee of Inquiry* (1964)

66 The Argentines had tried similar tactics with the US, to considerably lesser effect; unlike Britain, the US was capable of producing sufficient meat to feed the nation. Nevertheless, FMD had an important impact upon relations between the US and Latin American countries. See M Machado (1969).

67 Correspondence, January–February 1968, PRO MAF 287/522

68 Memos and correspondence, January 1968, PRO MAF 287/464 and MAF 287/465; *Origin of the 1967–1968 FMD epidemic* (1967–1968), pp227–34

69 Correspondence, February 1968, PRO MAF 287/465, MAF 287/466, MAF 287/520/1; NFU correspondence to prime minister, 12 February 1968, PRO PREM 13/1934

70 Correspondence and press cuttings, December 1967–February 1968, PRO FCO 7/168 and FCO 7/1070; M Gale correspondence, 22 and 23 February 1968, Gallagher to Sugg, 23 March 1968, PRO FCO 42/78. For details of Anglo–Argentine negotiations upon the power plant, see PRO AB 65/500 and AB 64/1107. Post-war Argentine governments were extremely supportive of the nuclear industry, believing that it would raise the domestic prestige of the government and the international standing of the nation as a whole. See D Poneman (1984), p878.

71 Cited in *Hansard* [HC], 30 January 1968, vol 757, col 1170. For further discussions, see *Hansard* [HC], 30 January 1968, vol 757, col 1134–90; ibid, 15 February, vol 758, col 1585–99.

72 Correspondence, February 1968, PRO MAF 287/465 and MAF 287/466

73 Correspondence, February 1968, PRO FCO 42/68, FCO 7/168, FCO 7/1070, PREM 13/1934, CAB 130/370

74 'Foot and mouth disease', *The Times,* London, 5 March 1968, p2, col a; ibid, p6, col b

75 *Hansard* [HC], 4 March, vol 760, col 40–49; ibid, 13 March, col 1388–1508

76 *Origin of the 1967–1968 FMD epidemic* (1967–1968), pp39, 227–34; Correspondence and reports, January–February 1968, PRO MAF 287/467

77 As Chapter 4 showed, during the 1920s and 1930s MAF had itself discounted the circumstantial link between Argentine meat imports and British FMD outbreaks, so one can hardly blame the Argentines for using this argument to their advantage. See comments by Argentine meat board and *Cronista,* PRO FCO 7/1070; *La fiebre aftosa en Gran Bretana anos 1967–1968,* FCO 7/71; Text of statement circulated in Argentine press, 12 March 1968, PRO MAF 287/520/1

78   Obituary: W Henderson (2000) 'Veterinary Public Health', http://165.
      158.1.110/english/pro_salute/history182.pdf, p5
79   Correspondence and 'Report: The FMD situation', PRO MAF 287/520/1
80   Cresswell to FO, 5 April 1968, PRO MAF 276/396
81   Gale to Diggines, 29 March 1968, PRO FCO 7/168; J Carnochan to W
      Tame, 5 August 1968, PRO MAF 287/529; M Cresswell, 'The Anglo–
      Argentine meat problem: summary', PRO MAF 287/520/1
82   *Cronista*, 3 April 1968, PRO FCO 7/1070. Houssay (1887–1971) was the
      most prominent Argentine physiologist of his generation. He overcame
      unfavourable working conditions and persecution under Peron to carry out
      research in endocrinology. He was well regarded in the West, and in 1947
      became the first Latin American to receive a Nobel Prize in science; M Cueto
      (1994)
83   'Now Britain gets blame for farm plague', *Daily Express*, London, 22 April
      1968, PRO MAF 276/396; British embassy, Buenos Aires to Diggines, 29
      April 1968, PRO FCO 7/1070. The Houssay mission's findings formed the
      basis for a government White Paper, *La fiebre aftosa en Gran Bretana anos
      1967–1968*, published in June 1968. See PRO FCO 7/171.
84   Correspondence, May–July 1968, PRO FCO 7/168 and PRO 7/1070; *La
      fiebre aftosa en Gran Bretana anos 1967–68*, PRO FCO 7/171
85   Correspondence, March–May 1968, PRO FCO 7/1070; 'Till the cows come
      home', *Economist*, London, 11 May 1968, pp67–70; Correspondence, June–
      July 1968, PRO BT 11/6959; Memo, 18 September 1968, PRO FCO 7/168;
      'The ban that became a boon', *Economist*, London, 25 January 1969, p69
86   Correspondence March–April 1968, PRO MAF 276/396; FO to British
      embassy, Buenos Aires, September 1968, 'Announcement of agreement', 10
      October 1968, PRO FCO 7/171; 'The ban that became a boon', *Economist*,
      London, 25 January 1969, p69

Post-mortem and aftermath

87   Mrs White to Miss A Hills, 16 May 1968, Carnochan minutes, 21 May 1968,
      PRO MAF 287/505/1; Meeting 10 July 1968, PRO MAF 287/520/1
88   Correspondence, PRO MAF 283/693, MAF 287/502, MAF 287/503 and
      MAF 287/504. Significantly, the published version of this paper discussed only
      slaughter and vaccination and excluded the politically sensitive issue of meat
      import policy. See A Power and S Harris (1973).
89   W Hughes to MAFF, 22 March 1968, PRO BT 11/6921; reports and corre-
      spondence, March 1968–May 1969, PRO MAF 283/693; correspondence,
      September 1968, PRO MAF 287/503.
90   'Effects on British trade with South America under alternative meat imports
      scheme', November 1968, PRO BT 11/6921
91   Report of the Committee of Inquiry into FMD (Northumberland Com-
      mittee), Part I (1968–69), p93

92 Argentine embassy and FO, November 1968, PRO FCO 7/1050; Correpond-
    ence, March–April 1969, PRO FCO 7/1094
93 Correspondence March–April 1969, PRO MAF 276/403; Hughes group
    report, 1 April 1969, PRO FCO 67/71; D Evans minutes, 28 April 1969,
    PRO MAF 276/403; Stewart to British embassy, Buenos Aires, 28 April 1969,
    EID memo, 'Argentina and the UK beef trade', 30 April 1969, Secretary of
    State to Richard North, MP, 7 May 1969, PRO FCO 7/1094; Discussions on
    meat import tariff, April–July 1969, PRO FCO 7/1050; Meat Division
    response to Board of Trade report, November 1968, PRO BT 11/6921
94 Like its predecessor, the Northumberland Committee of Inquiry concluded
    that the 'theoretical' risk of vaccinated animals becoming disease carriers had
    been exaggerated, and drew examples from Europe to show that vaccines were
    capable of controlling FMD. *Report of the Committee of Inquiry into FMD*
    (1968–1969), pp64–100
95 Hughes group report, 2 April 1969, PRO MAF 276/403; Carnochan to
    Treasury, 23 April 1969, Cabinet Ministerial Committee on Economic Policy,
    28 April 1969, PRO MAF 276/403; H G Button minutes, 25 April 1969,
    Report on meeting at Treasury, 1 May 1969, PRO MAF 287/479/1
96 Carnochan minutes, 15 September 1969, PRO MAF 287/479/1. The follow-
    ing documents also provide evidence of declining enthusiasm for vaccination:
    Report by officials' working party, 1 April 1968, FCO 67/71; the North-
    umberland Committee report, commercial policy implications, 2 April 1969,
    MAF 276/403; Draft Parliamentary statement, 1 May 1969, PREM 13/248
97 Reid to Carnochan, 3 October 1969, PRO MAF 287/479/1
98 'FMD – ring vaccination contingency plan, October 1969', PRO MAF 287/
    479/2.
99 The above two paragraphs are drawn from correspondence and memos, PRO
    MAF 287/479/1 and MAF 287/479/2; D Christie, L Reynolds and E Tansey
    (2003), pp67–68
100 The motivating force behind the ban was the French farm minister, Jacques
    Chirac. Peart, who had regained his post as British minister of agriculture, was
    extremely reluctant to support the proposal, but agreed to do so providing
    British beef farmers received additional EEC subsidies. *Economist*, London, 27
    April 1974, p106; ibid, 18 May 1974, p113; ibid, 6 July 1974, p63; ibid, 5
    October 1974, p93; ibid, 28 December 1974, p55; E Milenky (1978), pp31,
    137, 146

CHAPTER 8

Fighting FMD, 1968–2000

1 PRO MAF 35/765, MAF 35/868, MAF 35/869, MAF 252/48, MAF 252/499.
   The OIE was established by the French in 1924 in response to the renewed
   threat of cattle plague, which had appeared in Belgium in 1920. It was charged

with collecting and notifying governments of the facts pertaining to contagious animal diseases that called for international control efforts. It was also to carry out research into these diseases and to advise and assist governments in their execution of international agreements concerning veterinary police powers. See *Annual Report of Proceedings under the Diseases of Animals Acts* (1928), p4; 'A short history of the OIE', www.oie.int/eng/OIE/en_histoire.htm. The Anglo–French controversy over the formation of the EUFMD reflected and was influenced by the concurrent international political debates over the formation of the European Economic Community. See A Woods (2002), ch 7.

2   *Report of EUFMD* (1969 and 1970)
3   France and Belgium adopted annual mass vaccination in 1962; West Germany – where controls were previously left to individual states – adopted it in 1965; Luxembourg and Switzerland adopted it in 1966; Italy adopted it in 1968 and Spain in 1969. Anon (1978), p195
4   A Morrow, N Hyslop and L Buckley (1966), pp192–95; F Brown (2003), pp3–4
5   Rees (undated), unpublished, table 3
6   Anon (1978), p195
7   Rees (undated), unpublished
8   'Foot and mouth disease, UN Food and Agriculture Organization meeting plans', *The Times*, London, 17 July 1962, p10, col e; 'Foot and mouth disease: African type virus: spread reported', ibid, 18 October 1962, p10, col d; 'Foot and mouth disease, European spread checked by mass vaccination' ibid, 19 March 1963, p9, col b; Anon (1978), pp196–97
9   *Report of the EUFMD* (1977), appendix 7
10 *Report of the EUFMD* (1970); *Report of the EUFMD* (1979)
11 EEC Intra-Community Regulations, PRO MAF 287/490
12 MAFF communication to Northumberland Committee of Inquiry, 12 August 1968, PRO MAF 287/490
13 Rees (undated), unpublished
14 *Report of the EUFMD* (1981), p9 and appendix B6; *Report of the EUFMD* (1983)
15 *Report from the Commission on the control of FMD* (1989); Rees (undated) unpublished; G Davies (1993), pp1109–10; Y Leforban (1999), pp1755–56; I Anderson (2002)
16 Rees (undated), unpublished ; 'OIE official "disease–free" status', www.oie.int/eng/info/en_statut.htm; The Royal Society (2002), pp39–42
17 Obituary: W Henderson (2001)
18 'Veterinary Public Health', http://165.158.1.110/english/pro_salute/history182.pdf, p5
19 F Brown (1986), p220; P Kitching (1998), pp99–100; P Sutmoller et al (2003), pp127–30; 'Veterinary Public Health' http://165.158.1.110/english/pro_salute/history182.pdf, pp6–7; 'PAHO Epidemiological Bulletin, June 1998', www.paho.org/english/sha/epibul_95–98/EB_v19n2.pdf, pp14–16
20 P Kitching (1998), pp90–98
21 Y Leforban, (1999), pp1758–59

22   *European Commission Strategy for Emergency Vaccination against Foot and Mouth Disease* (1999)
23   J Ryan (2001), pp3–4
24   I Anderson (2002), pp34–36, 42

Epidemic, 2001

25   *Origin of the UK FMD epidemic in 2001* (2002), p12
26   'Ministry "failed to heed advice on pigswill"', *The Times*, London, 25 May 2001, p7; 'Farmer kept quiet about disease', *BBC News*, 30 May 2002; *Origin of the UK FMD epidemic in 2001* (2002); I Anderson (2002), pp45–49
27   Anderson (2002), pp49–51, 58–61; 'Quicker response would have halved FMD cull', *Daily Telegraph*, London, 16 July 2002
28   *Origin of the UK FMD epidemic in 2001* (2002), pp8–10
29   During the later 20th century, the FMD Research Institute at Pirbright extended its range of activities and began investigations into other infectious diseases of farm animals. It was later absorbed into the three-site Animal Health Institute.
30   This and the preceding two paragraphs are drawn from I Anderson (2002), pp66–80
31   A Richardson (2001); 'Brigadier's battle with No 10 over the FMD cull', *Daily Telegraph*, London, 18 February 2002; I Anderson (2002), pp26–28, 78–82, 88–96, 101–102
32   I Anderson (2002), pp115–19
33   '10 million animals were slaughtered in FMD cull', *Daily Telegraph*, London, 18 February 2002
34   'Green allies demand vaccination option', *Guardian*, London, 12 April 2001, p6; 'NFU "ignoring" small farmers', ibid, 19 April 2001, p8; 'Vets in open revolt on "needless slaughter"', *The Times*, London, 19 April 2001, p1; Roger Windsor, address to RCVS Council, 6 June 2001www.warmwell.com/windsor june18.htm; 'Lessons from an epidemic', *Nature*, 28 June 2001, p977; RG Eddy correspondence, *Nature*, 2 August 2001, p477; Vets for Vaccination meeting with Royal Society, January 2002, www.smallholders.org/FMD/ nfmgvfvrs230102.rtf
35   'Animal cull "based on incorrect assumption"', *Daily Telegraph*, London, 19 February 2002; Dr Paul Kitching, interview with Channel 4 news, 21 April 2001, www.farmtalking.com/news–ch4–kitching.htm; Various, 'The Small World of Professor Krebs and Professor Anderson', www.warmwell.com/ andersongroup.html
36   For a selection of arguments against MAFF policy, see House of Commons Research Paper 01/35 (2001); 'FMD cure "was worse than disease"', *Guardian*, London, 1 October 2001.
37   B Mepham (2001); N Morris, 'The reality of the pre-emptive cull in UK 2001 FMD epidemic', www.warmwell.com/july1lnicolamorriscull.htm; 'Farmers rebel as MAFF admits cull blunders', *Sunday Times*, London, 22 April 2001,

p28; 'Third of all "positive" cases are wrong', *Guardian*, London, 11 May 2001, p10; G Thomas-Everard, 'Briefing note to temporary committee on FMD', www.warmwell.com/ap15everard.html

38   P Midmore, 'Economic Arguments against an extended cull', www.efrc.com/fmd/fmdtext/fmdecon.pdf

39   FMD Lessons Learned Inquiry, meeting with National Foot and Mouth Group, www.warmwell.com/july4nfmgll.htm; P Sutmoller et al (2003)

40   'Talk of vaccination', *The Times*, London, 19 April 2001, p19; 'Propaganda war over vaccination continues', ibid, 21 April 2001, p14

41   'Farmers oppose expected U-turn on vaccination', *The Times*, London 17 April 2001, p8; 'Farmers will not back vaccination proposal', ibid, 18 April 2001, p6; A Kaletsky, 'What is so special about the farmers?', ibid, 19 April 2001, p18

42   See the various contributions to www.warmwell.com; L Purves, 'This sickening shambles must be stopped now', *The Times*, London, 17 April 2001, p16

43   See *The Times, Telegraph, Independent* and *Guardian*, February–June 2001, passim.

44   J Freedland, 'A catalogue of failures that discredits the whole system', *Guardian*, London, 16 May 2001, p22; 'Wasted nation: the truth about FMD', *The Times*, London, section 2, 24 May 2001, pp2–8; articles by Nick Green on www.warmwell.com

45   'Farmers deny FMD illegalities', *BBC News*, 10 April 2001; 'FMD', *The Times*, London, 12 April 2001, p25; I Anderson (2002), pp148–51

46   'Slaughter policy to be relaxed', *BBC News*, 26 April 2001; I Anderson (2002), p146; S Jenkins, 'You must vaccinate or be damned, Mr Gill', *The Times*, London, 1 August 2001, p14

47   'Drastic cull is only cure', *The Times*, London, 13 April 2001, p12; A Kirby, 'Foot and mouth: a pyrrhic victory?' *BBC News*, 3 May 2001; P Kitching, 'Submission to the temporary committee on FMD', http://www.warmwell.com/july20kitch.html

48   'New green ministry faces tests', *BBC News*, 11 June 2001

49   DEFRA (2001), 'Histogram of Confirmed Cases', www.defra.gov.uk/footandmouth/cases/histogram.htm

50   'Foot and mouth: Blair backs out of public inquiry', *Guardian*, London, 20 July 2001

51   'Ruling against public inquiry "was flawed"', *Daily Telegraph*, London, 19 February 2002; 'Whitehall nervous as calls for inquiry grow', *Guardian*, London, 19 February 2002; '250,000 in call for FMD inquiry', *Daily Telegraph*, London, 6 March 2002; 'Farmers lose bid for foot and mouth inquiry', *Guardian*, London, 15 March 2002

Conclusion

1   Woods (2004a)
2   I Anderson (2002); The Royal Society (2002); European Parliament (2002)
3   A Blake, M Sinclair, G Sugiarto, 'Quantifying the impact of FMD on Tourism and the UK economy', www.nottingham.ac.uk/~lizng/ttri/Pdf/FMD–paper4. PDF
4   National Audit Office (2002), p2
5   I Anderson (2002), p138
6   I Anderson (2002), pp6–19; The Royal Society (2002), ppvii–xiv, ch 8; European Parliament (2002), pp3, 9–11
7   I Anderson (2002), p6
8   I Anderson (2002), p13
9   The Royal Society (2002), pviii
10 DEFRA (2003), *Foot and Mouth Disease Contingency Plan*, www.defra.gov. uk/footandmouth/contingency/contplan.pdf
11 National Foot and Mouth Group (2003), 'Response to DEFRA's contingency plan', www.warmwell.com/nfmgresponseconting.html
12 D Campbell and R Lee (2003), p383

References

ARCHIVES AND UNPUBLISHED SOURCES

Public Records Office (National Archives)

Records of the UK Atomic Energy Authority	AB, series 64, 65
Board of Trade	BT 11
Cabinet Office	CAB 27, 124, 130
Development Commission	D 4
Ministry of Defence	DEFE 10
Foreign and Commonwealth Office	FCO 7, 23, 42, 67
Medical Research Council	FD1/1346, 1/4364, 1/5048
Ministry of Agriculture	MAF 33, 35, 85, 88, 117, 124, 246, 250, 252, 255, 276, 283, 287, 387, 394
Prime Minister's Office	PREM 13
Treasury	T 161
War Office	WO 195, 208, 291

Cheshire Records Office

Cheshire Farmers' Union minutes, 1923–1924
Cheshire County Council Diseases of Animals Committee minutes, 1923–1924

Institute for Animal Health, Pirbright

Minutes of FMD Research Committee Meetings (FMDRC CM), 1924–1952
FMD Research Committee Papers, (FMDRC CP), 1924–1952
Correspondence and unpublished reports relating to FMD research, 1924–1953
Collection of newspaper clippings from 1951–1952 FMD epidemic
H Skinner (198–), 'The British contribution to research on FMD prior to 1950', in ten sections

Rural History Centre, Reading University
NFU Meat and Livestock Committee minutes, 1919–1940
RAS Council and Veterinary Committee minutes, 1914–1953

Royal Veterinary College, London

J B Simonds Collection

Oral history

Mary Brancker, FRCVS, interview, 4 November 2002
Howard Rees, FRCVS, interview, December 2002
H H Skinner, FRCVS, interview, 7 March 2000
Professor A Steele-Bodger FRCVS, interview, 27 January 2003

Other unpublished sources

P Atkins, 'White Heat in Whitehall: Inter-departmental friction and its impact upon food safety policy: the example of milk, 1930–1935' (2002)
A Kraft 'Breaking with Tradition: The Reform of British Veterinary Education, 1900–1930' (2003)
J A Mendelsohn, *Cultures of Bacteriology: Formation and Transformation of a Science in France and Germany, 1870–1914* (PhD thesis, Princeton, 1996)
H Rees, 'FMD: its control and eradication. The European story' (date unknown)
A Woods, *From occupational hazard to animal plague: Foot-and-mouth disease in Britain, 1839–1884*, (MSc thesis, University of Manchester, 1999)
A Woods, *Foot and Mouth Disease in 20th Century Britain: Science, Politics and the Veterinary Profession*, (PhD thesis, University of Manchester, 2002)

NEWSPAPERS AND PERIODICALS

BBC News (www.news.bbc.co.uk)
British Medical Journal
Cheshire Observer
Cheshire Chronicle
Crewe Chronicle
Daily Mail
Economist
Guardian (www.guardian.co.uk/footandmouth)
Journal of the Board/Ministry of Agriculture
Lancet

Nature
Daily Telegraph (www.portal.telegraph.co.uk)
The Times
Veterinary Journal
Veterinary News
Veterinary Record

OFFICIAL PUBLICATIONS

Hansard

Parliamentary Debates
Parliamentary Debates: Lords

Bills

Bill to Extend the Provision and Continue the Term of the Act of the 12th Year to Prevent Spreading of Contagious and Infectious Disorders among Sheep and Cattle, PP, 1863, vol IV, 181

Bill to Make Further Provisions for the Prevention of Infectious Diseases amongst Cattle, PP, 1864, vol I, 127

Bill to Consolidate, Amend and Make Perpetual the Acts for Preventing the Introduction or Spreading of Contagious or Infectious Diseases among Cattle and Other Animals in Great Britain, PP, 1868–69, vol I, 519–60

Bill for Making Better Provision Respecting Contagious and Infectious Diseases of Cattle and other Animals, PP, 1878, vol I, 317

Bill for Making Better Provision Respecting Contagious and Infectious Diseases of Cattle and other Animals as Amended in Committee, PP, 1878, vol I, 423

Bill to Amend the Contagious Diseases (Animals) Act 1878, PP, 1884, vol XXI, 1

Reports

Third Report from the Select Committee on the Adulteration of Food, PP 1856, vol VIII, 1

Report from the Select Committee on the Sheep etc. Contagious Diseases Prevention Bill, PP 1857, vol IX, 649

E Headlam Greenhow, *Report on Murrain in Horned Cattle, the public sale of diseased animals and the effects of the consumption of their flesh on human health*, PP 1857 (session 2), vol XX, 367

Fifth Report of the Medical Officer of Health of the Privy Council, PP 1863, vol XXV, 1

J Gamgee, 'Cattle Diseases in Relation to the Supply of Meat and Milk', in *Fifth Report of the Medical Officer of Health of the Privy Council*, PP 1863, vol XXV, 1

Dr Edward Smith, 'Report on the food of the poorer labouring classes in England', in *6th Report of the Medical Officer of Health of the Privy Council*, PP 1864, vol XXVIII, 220–333

Report from the Select Committee on the Cattle Diseases Prevention and Cattle etc Importation Bills, PP 1864, vol VII, 235

Annual Report of the Agricultural Department (later renamed *Annual Reports of Proceedings under the Diseases of Animals Acts* and *Animal Health Services Report*), 1869–1969

Return of Number of Farms Infected with Foot-and-Mouth Disease, PP 1870, vol LVI, 663

Report of the Select Committee on the Contagious Diseases of Animals, PP 1873, vol XI, 189

Report of the Select Committee on Cattle Plague and the Importation of Livestock, PP 1877, vol IX, 1

Report of the Lords Select Committee on the Contagious Diseases of Animals Bill, PP 1878, vol XI, 71

Annual Report of the Department of Agriculture and Technical Instruction for Ireland, 1912–1916

Report of the 1912 Departmental Committee on FMD, PP 1912–13, cd 6222, Vol XXIX, 1

Evidence, Appendices and Index to *Report of the 1912 Departmental Committee on FMD*, PP 1912–13, cd 6244, Vol XXIX, 13

Report of the Departmental Committee Appointed to Inquire into the Requirements of the Public Services with Regard to Officers Possessing Veterinary Qualification, PP 1912–13, cmd 6575, xlviii, p251

Evidence, Appendices, and Index to *Report of the Departmental Committee Appointed to Inquire into the Requirements of the Public Services with Regard to Officers Possessing Veterinary Qualifications*, PP 1912–13, Cmd 6652, xlviii, p267

Department of Agriculture and Technical Instruction for Ireland, *Report on FMD in Ireland in the Year 1912*, PP 1914, cd 7103, xii, 793

Report of Proceedings at a Conference Held at Birkenhead on the 26th of February 1914, PP 1914, cd 7326, Vol XII, 171

Report of the Departmental Committee Appointed to Inquire into FMD, PP, 1914, cd 7270, Vol XII, 139

Report of the 1922 Departmental Committee on FMD, PP 1923, Cmd 1784, ii, 579

Report of the 1924 Departmental Committee on FMD, PP 1924–25, Cmd 2350, xiii, 225

1st–5th Progress Report of the FMD Research Committee, (1925, 1927, 1928, 1931, 1937), HMSO, London

Second Report of the Committee on Veterinary Education in Great Britain 1944 (HMSO, Cmd 6517),

FMD Research – Interim Report, (HMSO, London, 1952)

Report of the Departmental Committee on FMD, 1952–54, PP 1953–54, Cmd 9214, xiii, p561

Constitution of the European Commission for the Control of Foot-and-Mouth Disease, Rome, 11 December, 1953, PP 1953–54, Cmd 9283, xxxi, p539

Report of the ARC, 1956–57, PP 1957–58, Cmd 432, vii, p165

Reports of the 3rd – 24th Sessions of the European Commission for the Control of FMD (FAO, Rome, 1956–1981)

The Aberdeen Typhoid Outbreak of 1964: Report of the Departmental Committee of Inquiry (HMSO, 1964)

La fiebre aftosa en Gran Bretana, anos 1967–68 (Buenos Aires, 1968)

Origin of the 1967–68 Foot-and-Mouth Disease Epidemic, PP 1967–68, Cmd 3560, xxxix, p227

Report of the Committee of Inquiry into FMD (Northumberland Committee), Part One, PP 1968–69, Cmd 3999, xxx, p867

Report of the Committee of Inquiry into FMD (Northumberland Committee), Part Two, PP 1969–70, Cmd 4225, v, p157

Report from the Commission to the Council on a Study Carried out by the Commission on Policies Currently Applied by Member States in the Control of FMD (Commission of the European Communities, Brussels, 1989)

European Commission Strategy for Emergency Vaccination against Foot and Mouth Disease – Report of the Scientific Committee on Animal Health and Animal Welfare (1999), www.europa.eu.int/comm/food/fs/sc/scah/out22_en.html

Origin of the UK FMD Epidemic in 2001 (HMSO, London, 2002)

National Audit Office, *Report upon the 2001 FMD Outbreak* (London, HMSO, 2002)

I Anderson, *FMD 2001: Lessons to be Learned Enquiry Report* (London, HMSO, 2002)

The Royal Society, *Infectious Diseases in Livestock* (Royal Society, London, 2002)

European Parliament, *Report on Measures to Control FMD in the European Union in 2001* (2002), www2.europarl.eu.int/omk/sipade2?L=EN&OBJID=9904& LEVEL=3&MODE=SIP&NAV+X&LSTDOC+N

DEFRA, *Foot and Mouth Disease Contingency Plan* (2003), www.defra.gov.uk/footandmouth/contingency/contplan.pdf

BOOKS AND JOURNAL ARTICLES

F Accum (1820) *Treatise on Adulterations of Food*, J Mallett, London

E Ackerknecht (1948) 'Anticontagionism between 1821 and 1867', *Bulletin of the History of Medicine*, vol 22, pp562–93

The Analytical Sanitary Commission (1851) 'Milk and its adulterations', *Lancet* II, pp 257–262, 279–282, 322–325

K Angus (1990) *A History of the Animal Diseases Research Association*, ADRA, Edinburgh

Anon (1830) *Deadly Adulteration and Slow Poisoning Unmasked*, Sherwood, Gilbert and Piper, London

Anon (1840) 'The epidemic among cattle', *Journal of the Royal Agricultural Society of England*, vol 1, ppcxii–cxcvi

Anon (1965) *Animal Health: A Centenary*, HMSO, London

Anon (1967) 'The Central Veterinary Laboratory Weybridge, 1917–67', *Veterinary Record,* vol 81, pp62–68

Anon (1978) 'FMD', *Veterinary Record,* vol 102, pp184–98

J Austoker (1988) *The History of the Imperial Cancer Research Fund, 1902–1986,* Oxford University Press, Oxford

P Baldwin (1999), *Contagion and the State in Europe, 1830–1930,* Cambridge University Press, Cambridge

B Balmer (1997) 'The drift of biological weapons policy in the UK, 1945–65', *Journal of Strategic Studies,* vol 20, pp115–45

B Balmer and G Carter (1999) 'Chemical and biological warfare and defence, 1945–90' in R Bud and P Gummett (eds) *Cold War, Hot Science,* Harwood, Amsterdam

J M Beattie and D Peden (1924) 'FMD in rats', *Lancet,* 2 February, pp221–22

B Bernstein (1987) 'Churchill's secret biological weapons', *Bulletin of the Atomic Scientists,* vol 43, pp46–50

P Bew (1987) *Conflict and Conciliation in Ireland, 1890–1910,* Clarendon Press, Oxford

J Blackman (1975) 'The cattle trade and agrarian change on the eve of the railway age', *Agricultural History Review,* vol 23, pp48–62

D C Blood and O Radostits (1994) *Veterinary Medicine,* 8th edition, Bailliere, Tindall and Cox, London

C Booth (1987) 'Clinical research and the MRC' in C Booth (ed) *Doctors in Science and Society,* Memoir Club, London

C Booth (1993) 'Clinical research' in W Bynum and R Porter (eds) *Companion Encyclopaedia to the History of Medicine,* Routledge, London

J Bowers (1985) 'British agricultural policy since the second world war', *Agricultural History Review,* vol 33, pp66–76

M Brancker (1972) *All Creatures Great and Small: Veterinary Surgery,* Educational Explorers, Reading

P Brassley (2000a) 'Farming systems' in E Collins (ed) *The Agrarian History of England and Wales,* vol VII, part 1, Cambridge University Press, Cambridge

P Brassley (2000b) 'Output and technical change in twentieth-century British agriculture', *Agricultural History Review,* vol 48, pp60–84

J Broad (1983) 'Cattle plague in eighteenth-century England', *Agricultural History Review,* vol 31, pp104–15

F Brown (1986) 'Foot-and-mouth disease – one of the remaining great plagues', *Proceedings of the Royal Society London,* B229, pp215–26

F Brown (2003) 'The history of research into foot and mouth disease', *Virus Research,* vol 91, pp3–7

G Brown (1873) 'On the foot-and-mouth complaint of cattle and other animals', *Journal of the Royal Agricultural Society of England,* vol 34, pp437–82

J Brown (1987) *Agriculture in England, a Survey of Farming, 1870–1947,* Manchester University Press, Manchester

J Brown (2000) 'Agricultural policy and the NFU, 1908–1939' in J Wordie (ed) *Agriculture and Politics in England, 1815–1939,* Macmillan, Basingstoke, pp178–98

L Bryder (1999) '"We will not find salvation in inoculation": BCG vaccination in Scandinavia, Britain and the USA, 1921–1960', *Social Science and Medicine*, vol 49, pp1157–67

F Bullock (1930) *Handbook for Veterinary Surgeons*, Bailliere, Tindall and Cox, London, 2nd edition

J Burdon-Sanderson (1877), 'Report on the progress of investigations into the nature of pleuro-pneumonia and FMD', *Journal of the Royal Agricultural Society of England*, vol 38, pp204–07

J Burnett (1979) *Plenty and Want: A Social History of Diet in England from 1815 to the Present Day*, Scolar Press, London

D Campbell and R Lee (2003) 'The power to panic: the Animal Health Act 2002', *Public Law*, pp382–96

D Cannadine (1978) *The Decline and Fall of the British Aristocracy*, Papermac, London

F Capie (1978) 'Australia and New Zealand competition in the British market, 1920–39', *Australian Economic History Review*, vol 18, pp46–63

F Capie (1981) 'Invisible barriers to trade: Britain and Argentina in the 1920s', *Inter-American Economic Affairs*, vol 35, pp91–96

G Carter and G Pearson (1996) 'North Atlantic chemical and biological research collaboration, 1916–95', *Journal of Strategic Studies*, vol 19, pp74–103

D Cavallo and U Mundlak (1982) *Agricultural and Economic Growth in an Open Economy: The Case of Argentina*, International Food Policy Unit, Washington, DC

H Chick, M Hume and M MacFarlane (1971) *War on Disease – A History of the Lister Institute*, Andre Deutsch, London

D Christie, L Reynolds and E Tansey (2003) (eds) 'Wellcome witnesses to 20th century medicine', *Foot and Mouth Disease: The 1967 Outbreak and its Aftermath*, Wellcome Trust, London

F Clater (1853) *Every Man His Own Cattle Doctor*, Cradock and Co, London, 11th edition

G H Collinge, T Dunlop Young and A P McDougall (1929) *The Retail Meat Trade*, Gresham Publishing, London

A Cooper (1989) *British Agricultural Policy, 1912–36: A Study in Conservative Politics*, Manchester University Press, Manchester

G Cosco and A Aguzzi (1919) 'Virulence of the blood in FMD and attempts at immunisation', *Veterinary Review*, vol 3, pp28–29

G Cox, P Lowe and M Winter (1986) 'From state direction to self regulation: the historical development of corporatism in British agriculture', *Policy and Politics*, vol 14, pp475–90

G Cox, P Lowe and M Winter (1991) 'The origins and early development of the NFU', *Agricultural History Review*, vol 39, pp30—47

W M Crofton (1936) *The True Nature of Viruses*, London, J Bale

M Cueto (1994) 'Laboratory styles in Argentine physiology', *ISIS*, vol 85, pp228–46

A Cunningham (1992) 'Transforming plague: the laboratory and the identity of infectious disease' in A Cunningham and P Williams (eds) *The Laboratory Revolution in Medicine*, Cambridge University Press, Cambridge

L P Curtis (1968) *Anglo-Saxons and Celts: A Study of Anglo–Irish Prejudice in Victorian Britain*, New York University Press, New York

L P Curtis (1997) *Apes and Angels: The Irishmen in Victorian Caricature*, Institute Press, Washington, DC, Revised Edition

M Dando (1994) *Biological Warfare in the 21st Century*, Brasseys, London

R D'Arcy Thompson (1974) *The Remarkable Gamgees: A Story of Achievement*, Ramsay Head Press, Edinburgh

G Davies (1993) 'Risk assessment in practice: a foot and mouth disease control strategy for the European Community', *Revue Scientifique et Technique, Office International des Épizooties*, vol 12, pp1109–19

T DeJager (1993) 'Pure science and practical interests: the origins of the ARC, 1930–37', *Minerva*, vol 13, pp129–51

G M Dillon (1989) *The Falklands: Politics and War*, MacMillan, London

E Driver (1909), *Cheshire: Its Cheese-Makers, Their Homes, Landlords, and Supporters*, Bradford, Derwent

W Duguid (1877) 'Further notes of experiments at the Brown Institution on the communication of FMD', *Journal of the Royal Agricultural Society of England*, vol 38, pp460–62

M Durey (1979) *The Return of the Plague*, Humanities Press, Dublin

J Eyler (1987) 'Scarlet fever and confinement: the Edwardian debate over isolation hospitals', *Bulletin of the History of Medicine*, vol 61, pp1–24

F Fenner and E P Gibbs (1993) *Veterinary Virology*, 2nd edition, Academic Press Ltd, London

J Fisher (1979–1980) 'Professor Gamgee and the farmers', *Veterinary History*, vol 1, pp47–64

J Fisher (1980) 'The economic effects of cattle disease in Britain and its containment, 1850–1900', *Agricultural History*, vol 52, pp278–94

J Fisher (1993a) 'Not quite a profession: the aspirations of veterinary surgeons in England in the mid-nineteenth century', *Historical Research*, vol 66, pp284–302

J Fisher (1993b) 'British physicians, medical science and the cattle plague, 1865–66', *Bulletin of the History of Medicine*, vol 67, pp651–69

J Fisher (2000) 'Agrarian politics', in E Collins (ed) *The Agrarian History of England and Wales*, vol VII, part 1, Cambridge University Press, Cambridge

J Fisher (2003) 'To kill or not to kill: the eradication of contagious bovine pleuropneumonia in western Europe', *Medical History*, vol 47, pp314–31

G Fleming (1869) 'Eczema epizootica', *Veterinarian*, vol 42, pp881–94

G Fleming (1880) 'The spontaneous generation of contagious diseases', *Veterinarian*, vol 53, pp500–09

F Floud (1927) *The Ministry of Agriculture and Fisheries*, GP Putnams, London

M Foran (1998) 'The politics of animal health: the British embargo on Canadian cattle, 1892–1932', *Prairie Forum*, vol 23, pp1–17

W Foster (1983) *Pathology as a Profession in Great Britain and the Early History of the Royal College of Pathologists*, RCP, London

I Galloway (1952) 'FMD', *British Agricultural Bulletin*, vol 5, p67

S Gilley (1978) 'English attitudes to the Irish in England, 1780–1900', in C Holmes (ed) *Immigrants and Minorities in British Society*, George Allen, London

N Goddard (1988) *Harvests of Change: The RASE 1838–1988*, Quiller, London

P Goodwin (1981) 'Anglo–Argentine commercial relations: a private sector view, 1922–43', *Hispanic American History Review*, vol 61, pp29–51

R Gravil (1970) 'State intervention in Argentina's export trade between the wars', *Journal of Latin American Studies*, vol 2, pp147–73

D Grigg (1987) 'Farm size in England and Wales from early victorian times to the present', *Agricultural History Review*, vol 35, pp179–89

S Hall (1962) 'The cattle plague of 1865', *Medical History*, vol 6, pp45–58

C Hamlin (1992) 'Predisposing causes and public health in early nineteenth century medical thought', *Social History of Medicine*, vol 5, pp43–70

C Hamlin (1994) 'State medicine in Great Britain', in D Porter (ed) *The History of Public Health and the Modern State*, Rodopi, Amsterdam

A Hardy (1993a) 'Cholera, quarantine and the English preventive system, 1850–1895', *Medical History*, vol 37, pp250–69

A Hardy (1993b) *The Epidemic Streets: Infectious Disease and the Rise of Preventive Medicine, 1856–1900*, Clarendon Press, Oxford

A Hardy (1999) 'Food, hygiene and the laboratory: a short history of food poisoning in Britain, *circa* 1850–1950', *Social History of Medicine*, vol 12, pp293–311

A Hardy (2003) 'Professional advantage and public health: British veterinarians and state veterinary services', *Twentieth Century British History*, vol 14, pp1–23

M Harrison (1996) 'A question of locality: the identity of cholera in British India, 1860–1890', in D Arnold (ed) *Warm Climes and Western Medicine: The Emergence of Tropical Medicine, 1500–1900*, Rodopi Press, Amsterdam

W Henderson (1985) *A Personal History of the Testing of FMD Vaccines in Cattle*, Massey-Ferguson, Coventry

E P Hennock (1998) 'Vaccination policy against smallpox, 1835–1914: a comparison of England with Prussia and imperial Germany', *Social History of Medicine*, vol 11, pp49–71

E Higgs (2000) 'Medical statistics, patronage and the state: the development of the MRC Statistical Unit, 1911–48', *Medical History*, vol 44, pp323–40

B A Holderness (2000) 'Dairying' and 'Intensive livestock keeping', in E Collins (ed) *The Agrarian History of England and Wales*, vol VII, part 1, Cambridge University Press, Cambridge

J Homfray (1884) *A Practical Revival of Agriculture*, Kegan Paul, London

J Howard (1886) 'Foot-and-mouth disease: its history and teachings', *Journal of the Royal Agricultural Society of England*, vol 22, pp1–18

H Hughes and J Jones (1969) *Plague on the Cheshire Plain*, Dobson, London

H Jenkins (1873) 'Report on the trade in animals and its influence on the spread of foot and mouth', *Journal of the Royal Agricultural Society of England*, vol 34, pp187–245

T A Jenkins (1996) *Parliament, Party and Politics in Victorian Britain*, Manchester University Press, Manchester

D Jones (1983) 'The cleavage between graziers and peasants in the land struggle, 1890–1910', in S Clark and J Donnelly (eds) *Irish Peasants: Violence and Political Unrest, 1780–1914*, Manchester University Press, Manchester

H Kamminga and A Cunningham (1995) 'Introduction', in H Kamminga and A Cunningham (eds) *The Science and Culture of Nutrition, 1840–1940*, Rodopi, Amsterdam

H Keary (1948) 'Management of cattle', *JRASE*, vol 9, p446

P Kitching (1998) 'A recent history of FMD', *Journal of Comparative Pathology*, vol 118, pp89–108

P de Kruif (1926) *Microbe Hunters*, Harcourt Brace and Co, New York

A Landsborough Thomson (1973) *Half a Century of Medical Research: Volume 1 – Origins and Policy of the Medical Research Council (UK)*, HMSO, London

B Latour (1988) *The Pasteurization of France*, translated by A Sheridan and J Law, Harvard University Press, London

C Lawrence (1985) 'Incommunicable knowledge: science, technology and the clinical art in Britain, 1850–1914', *Journal of Contemporary History*, vol 20, pp503–20

C Lawrence (1998) 'Still incommunicable: clinical holists and medical knowledge in inter-war Britain', in C Lawrence and G Weisz (eds) *Greater than the Parts: Holism in Biomedicine, 1920–50*, Oxford University Press, Oxford

Y Leforban (1999) 'Prevention measures against FMD in Europe in recent years', *Vaccine*, vol 17, pp1755–59

L Lomnitz and L Mayer (1994) 'Veterinary medicine and animal husbandry in Mexico: from empiricism to science and technology', *Minerva*, vol 32, pp144–57

O MacDonagh (1977) *Early Victorian Government 1830–1870*, Weidenfeld and Nicolson, London

G MacDonald (1980) *One Hundred Years – Wellcome: 1880–1980. In Pursuit of Excellence*, Wellcome Foundation, London

M Machado (1968) *An Industry in Crisis, Mexican–US Co-operation in the Control of FMD*, Berkeley, University of California Press

M Machado (1969) *Aftosa: A Historical Survey of FMD and Inter-American Relations*, University of New York Press, New York

E Madden (1984) *Brucellosis: A History of the Disease and Its Eradication from Great Britain*, MAFF, London

K Maglen (2002) '"The first line of defence": British quarantine and the port sanitary authorities in the nineteenth century', *Social History of Medicine*, vol 15, pp413–28

J Martin (2000) *The Development of Modern Agriculture: British Farming since 1931*, Macmillan, Basingstoke

A H H Matthews (1915) *Fifty Years of Agricultural Politics: Being the History of the Central Chamber of Agriculture*, P S King and Son, London

J C McDonald (1951) 'The history of quarantine in Britain during the 19th century', *Bulletin of the History of Medicine*, vol 25, pp22–44

(1953) 'Memoir', J T G Edwards, *British Veterinary Journal*, vol 109, pp76–84

(1952) 'Memoir, obituary and tributes to Henry William Steele-Bodger', *Veterinary Record*, vol 44, pp44–45, 56–59

B Mepham (2001) 'Foot and mouth disease and British agriculture, ethics in a crisis', *Journal of Agricultural and Environmental Ethics*, vol 14, pp339–47

W Mercer (1963) *A Survey of the Agriculture of Cheshire*, RAS, London
E Milenky (1978) *Argentina's Foreign Policies*, Westview, Colorado
K Miller (1985) *Emigrants and Exiles: Ireland and the Irish Exodus to North America*, Oxford University Press, Oxford
J Mitchell (1848) *A Treatise on the Falsifications of Food and the Chemical Means Employed to Unmask Them*, Balliere, London
S Moore (1991) 'The agrarian Conservative party in Parliament, 1920–29', *Parliamentary History*, vol 10, pp342–62
S Moore (1993) 'The real "great betrayal"? Britain and the Canadian cattle crisis of 1922', *Agricultural History Review*, vol 41, pp155–68
R Morris (1976) *Cholera 1832: The Social Response to an Epidemic*, Croom Helm, London
A Morrow, N Hyslop and L Buckley (1966) 'Formalin inactivated FMD vaccines prepared on an industrial scale, Part I – Production', *Veterinary Record*, vol 78(1), pp2–7
B Nerlich, C Hamilton, and V Rowe (2001) 'Conceptualising Foot and Mouth Disease: The Socio-Cultural Role of Metaphors, Frames and Narratives', www.metaphorik.de/02/nerlich.htm
'Obituary: Professor J Beattie' (1955) *Lancet*, 22 October, pp880–81
'Obituary: Sir W M Fletcher, 1873–1933' (1932–1935) *Royal Society Obituary Notices of Fellows*, vol 1, pp153–63
'Obituary: Sir William Leishman' (1995) *Dictionary of National Biography*, Oxford University Press, Oxford
'Obituary: W Henderson' (2000) *The Guardian*, 19 December, p20
'Obituary: Sir John McFadyean' (1995), *Dictionary of National Biography*, Oxford University Press, Oxford
'Obituary: Stewart Stockman' (1926) *The Times*, 4 June 1926, p19, col a
'Obituary: Dr W R Wooldridge' (1966) *Veterinary Record*, vol 79, pp314–20
A O'Day (1998) *Irish Home Rule, 1867–1921*, Manchester University Press, Manchester
R Olby (1991) 'Social imperialism and state support for agricultural research in Edwardian Britain', *Annals of Science*, vol 48, pp509–26
I Pattison (1979) 'John McFadyean and Stewart Stockman', *Veterinary History*, vol 1, pp2–10
I Pattison (1981) *John McFadyean, Founder of Modern Veterinary Research*, J A Allen, London
I Pattison (1984) *The British Veterinary Profession*, J A Allen, London
I Pattison (1990) *A Great British Veterinarian Forgotten; J B Simonds*, J A Allen, London
M Pelling (1978) *Cholera, Fever and English Medicine 1825–1865*, Oxford University Press, Oxford
R Perren (1978) *The Meat Trade in Britain, 1840–1914*, Routledge, London
P J Perry (1973) 'Introduction', in P J Perry (ed) *British Agriculture, 1875–1914*, Methuen, London
J Pickstone (1992) 'Dearth, dirt and fever epidemics: rewriting the history of British public health', in T Ranger and P Slack (eds) *Epidemics and Ideas*, Cambridge University Press, Cambridge

D Poneman (1984) 'Nuclear proliferation prospects for Argentina', *Orbis,* vol 24, pp853–80

A Power and S Harris (1973) 'A cost–benefit evaluation of alternative disease control policies for FMD in Great Britain', *Journal of Agricultural Economics,* vol 24, pp573–600

A Rabinbach (1992) *The Human Motor: Energy, Fatigue and the Origins of Modernity,* University of California Press, Berkeley

L Randall (1978) *An Economic History of Argentina in the 20th Century,* Columbia University Press, New York

J Richelet (1929) *The Argentine Meat Trade,* Sté Industrielle d'Imprimerie, London

H Ritvo (1987) *The Animal Estate,* Harvard University Press, London

D Rock (1985) *Argentina 1516–1982,* University of California Press, Berkeley

T Romano (1997) 'The cattle plague of 1865 and the reception of the germ theory', *Journal of Medical History,* vol 52, pp51–80

T Romano (2002) *Making Medicine Scientific: John Burdon Sanderson and the Culture of Victorian Science,* John Hopkins University Press, Baltimore

T Rooth (1985) 'Trade agreements and the evolution of British agricultural policy in the 1930s', *Agricultural History Review,* vol 33, pp173–90

C Rosenberg (1992) 'Framing disease', 'What is an epidemic? AIDS in historical perspective' and 'Explaining epidemics', in *Explaining Epidemics,* Cambridge University Press, Cambridge

P J Rowlinson (1982) 'Food adulteration – its control in 19th-century Britain', *Interdisciplinary Science Reviews,* vol 7, pp63–71

G Scard (1981) 'Squire and tenant: rural life in Cheshire 1760–1900', vol 10 of J Bagley (ed) *A History of Cheshire,* Cheshire Community Council, Chester

C Schenk (1998) 'Austerity and boom', in P Johnson (ed) *Twentieth Century Britain: Economic, Social and Cultural Change,* Longman, London

H P Schmiedebach (1999) 'The Prussian state and microbiological research – Friedrich Loeffler and his approach to the "invisible" virus' in C Calisher and M Horzineck (eds) *100 Years of Virology – The Birth and Growth of a Discipline,* SpringerWienNewYork, Austria

P Self and H Storing (1963) *The State and the Farmer,* University of California Press, Berkeley

W Sewell (1841) 'Report on the epidemic among cattle', *Journal of the Royal Agricultural Society of England,* vol 2, ppcxic–cxxi

D Sheinin (1994) 'Defying infection: Argentine FMD policy 1900–1930', *Canadian Journal of History,* vol 29, pp501–23

S E D Shortt (1983) 'Physicians, science, and status: issues in the professionalisation of Anglo–American medicine in the 19th century', *Medical History,* vol 27, pp51–68

H Skinner (1989) 'The origins of virus research at Pirbright', *Veterinary History,* vol 6, pp31–40

H Skinner (1992) 'The 1937 study tour in Germany by students of the RVC, London', *Veterinary History,* vol 7, pp15–18

H Skinner (1993) 'The 1951–52 FMD outbreaks in the UK: lay criticisms of the control measures and research and the official response', *Veterinary History,* vol 7, pp110–24

F Smith (1933) *The Early History of Veterinary Literature*, J A Allen, London

M Smith (1990) *The Politics of Agricultural Support in Britain*, Aldershot, Dartmouth

P Smith (1969) *Politics and Beef in Argentina*, Columbia University Press, London

S Smith Hughes (1977) *The Virus – A History of the Concept*, Heinemann Educational, London

C Solberg (1971) 'Rural unrest and agrarian policy in Argentina, 1912–30', *Journal of Inter-American Studies and World Affairs,* vol 13, pp15–55

D Spear (1982) 'California besieged: the foot-and-mouth epidemic of 1924', *Agricultural History,* vol 56, pp528–41

D Spring (1984) 'Land and politics in Edwardian England', *Agricultural History,* vol 58, pp17–42

S Stockman (1924) 'Address to the Council of the National Veterinary Medical Society', *Veterinary Record,* 19 January 1924, pp35–39

S Sturdy and R Cooter (1998) 'Science, scientific management and the transformation of medicine in Britain, c1870–1950', *History of Science,* vol xxxvi, pp421–66

P Sutmoller, S Barteling, R Olascoaga and K Sumption (2003) 'Review: control and eradication of foot and mouth disease', *Virus Research,* vol 91, pp101–44

E Tansey (1994) 'Protection against dog distemper and dogs protection bills: the Medical Research Council and anti-vivisectionist protest, 1911–1933', *Medical History,* vol 38, pp1–26

D Taylor (1975) 'The importance of cattle enterprises in British agriculture', *Veterinary History,* vol 5, pp19–32

D Taylor (1987) 'Growth and structural change in the English dairy industry, c1860–1930', *Agricultural History Review,* vol 35, pp47–64

J Thompson (1970) 'Argentine economic policy under the Ongania regime', *Inter-American Economic Affairs,* vol 24, pp51–75

M Tracy (1989) *Government and Agriculture in Western Europe, 1880–1988,* Harvester, London, 3rd Edition

T van Helvoort (1994) 'History of virus research in the twentieth century: the problem of conceptual continuity', *History of Science,* vol 32, pp186–235

K Vernon (1990) 'Pus, sewage, beer and milk: microbiology in Britain, 1870–1940', *History of Science,* vol xxviii, pp289–325

K Vernon (1997) 'Science for the farmer? Agricultural research in England, 1909–1936,' *20th Century British History,* vol 83, pp310–33

K Waddington (2001) 'A science of cows: tuberculosis, research and the state in the UK, 1890–1914', *History of Science,* vol xxxix, pp355–81

J Walton (1986) 'Pedigree and the national cattle herd circa 1750–1950', *Agricultural History Review,* vol 34, pp149–70

J H Warner (1995) 'The history of science and the sciences of medicine', *Osiris,* vol x, pp164–93

A P Waterson and L Wilkinson (1978) *An Introduction to the History of Virology,* Cambridge, Cambridge University Press

C Watson (1984) 'Will Argentina go to the bomb after the Falklands?' *Inter-American Affairs,* vol 37, pp63–80

P Weindling (1992) 'Scientific elites and laboratory organisation in fin de siècle Paris and Berlin' in A Cunningham and P Williams (eds) *The Laboratory Revolution in Medicine*, Cambridge University Press, Cambridge

P Weindling (1993) 'The immunological tradition', in W Bynum and R Porter (eds) *Companion Encyclopaedia of the History of Medicine*, Routledge, London

E Whetham (1979) 'The trade in pedigree livestock 1850–1910', *Agricultural History Review*, vol 27, pp47–50

R Whitlock (1969) *The Great Cattle Plague*, Country Book Club, London

G Williams and J Ramsden (1990) *Ruling Britannia: A Political History of Britain, 1688–1988*, Longman, London

W Wilson and R C Matheson (1952–1953) 'Bird migration and FMD', *Journal of the Ministry of Agriculture*, vol 59, pp213–28

J Winnifrith (1962) *The Ministry of Agriculture, Fisheries and Food*, George Allen and Unwin, London

W Wittmann (1999) 'The legacy of Friedrich Loeffler – the Institute on the Island of Riems' in C Calisher and M Horzineck (eds) *100 Years of Virology – The Birth and Growth of a Discipline*, SpringerWienNewYork, Austria

A Woods (2004a) 'The contruction of an animal plague: foot and mouth disease in nineteenth century Britain', *Social History of Medicine*, vol 17, pp23–39

A Woods (2004b) 'Fear and flames on the farm: controlling foot and mouth disease in Britain, 1892–2001, *Historical Research*, vol 77

W Wooldridge (1954) *Farm Animals in Health and Disease*, Crosby, Lockwood and Son, London

M Worboys (1991) 'Germ theories of disease and British veterinary medicine, 1860–1900', *Medical History*, vol 35, pp308–27

M Worboys (1992) 'Vaccine therapy and laboratory medicine in Edwardian Britain', in J Pickstone (ed) *Medical Innovations in Historical Perspective*, Macmillan, Basingstoke

M Worboys (2000) *Spreading Germs: Disease Theories and Medical Practices in Britain, 1860–1900*, Cambridge University Press, Cambridge

G Wynia (1978) *Argentina in the Postwar Era*, University of New Mexico Press, Albuquerque

WEB SOURCES

DEFRA, 'Histogram of confirmed cases' (2001), www.defra.gov.uk/footand mouth/cases/histogram.htm

FMD: House of Commons research paper 01/35 (27 March 2001), www.parliament.uk/commons/lib/research/rp2001/rp01–035.pdf

FMD Lessons Learned Inquiry, meeting with National Foot and Mouth Group (2002), www.warmwell.com/july4nfmgll.htm

'History of the Veterinary Laboratories Agency', www.defra.gov.uk/corporate/vla/aboutus/aboutus–history.htm

'A short history of the OIE', www.oie.int/eng/OIE/en_histoire.htm

'OIE official "disease-free" status', www.oie.int/eng/info/en_statut.htm

'PAHO Epidemiological Bulletin, June 1998', www.paho.org/english/sha/epibul_ 95–98/EB_v19n2.pdf

Dr Paul Kitching, interview with Channel 4 news (21 April 2001), www.farm talking.com/news–ch4–kitching.htm

P Kitching, 'Submission to the temporary committee on FMD' (2002), www. warmwell.com/july20kitch.html

P Midmore, 'Economic arguments against an extended cull' (2001), www.efrc.com/ fmd/fmdtext/fmdecon.pdf

N Morris, 'The reality of the pre-emptive cull in UK 2001 FMD epidemic' (2001), www.warmwell.com/july11nicolamorriscull.htm

A Richardson, 'Observations on the FMD outbreak in Cumbria, 2001' (2001), www.humanitarian.net/biodefense/papers/arofmd.doc

J Ryan, 'FMD, risk and Europe' (2001), www.cmlag.fgov.be/eng/JohnRyanEC.pdf

G Thomas-Everard, 'Briefing note to temporary committee on FMD' (2001), www.warmwell.com/ap15everard.html

Various (2001), 'The small world of Professor Krebs and Professor Anderson', www.warmwell.com/andersongroup.html

Various, www.warmwell.com

'Veterinary public health', http://165.158.1.110/english/pro_salute/history182.pdf

Vets for Vaccination meeting with Royal Society (January 2002), www. smallholders.org/FMD/nfmgvfvrs230102.rtf

'The small world of Professor Krebs and Professor Anderson' (2001), www. warmwell.com/andersongroup.html

Roger Windsor, address to RCVS Council (6 June 2001), www.warmwell.com/ windsorjune18.htm

Index